T0017414

VOYAGER

Also by Nona Fernández

The Twilight Zone
Space Invaders

VOYAGER

◄ Constellations of Memory ►

Nona Fernández

Translated from the Spanish by Natasha Wimmer

Graywolf Press

This publication is made possible, in part, by the voters of Minnesota through a
Minnesota State Arts Board Operating Support grant, thanks to a legislative
appropriation from the arts and cultural heritage fund. Significant support has
also been provided by the McKnight Foundation, the Lannan Foundation, the
Amazon Literary Partnership, and other generous contributions from foundations, corporations, and individuals. To these organizations and individuals
we offer our heartfelt thanks.

Published by Graywolf Press
212 Third Avenue North, Suite 485
Minneapolis, Minnesota 55401

www.graywolfpress.org

Published in the United States of America

ISBN 978-1-64445-217-2 (paperback)
ISBN 978-1-64445-218-9 (ebook)

2 4 6 8 9 7 5 3 1
First Graywolf Printing, 2023

Library of Congress Control Number: 2022938631

Cover design: Kapo Ng

For Patricia,
the mother star

The *Voyagers* are two space probes launched by NASA in 1977. Multiple arms and antennae jut from them, making them look curiously like cosmic insects. Their sophisticated framework accommodates cameras, light sensors, sonic radar: an array of instruments for measuring and gauging temperature, color, plasma waves, particle energy. The *Voyagers* are equipped to be two perfect huntresses. Their task is to record. To store fragments of stellar memory.

We must bring new energy to remembering.
Make memory talk to the troubled present.

—Nelly Richard

While I live, I remember.

—Agnès Varda

VOYAGER

Southern Cross

My mother has been fainting. Without warning, for no ap-
parent reason, she falls and briefly disconnects. It might be
a few minutes or only a few seconds, but when she comes to
she can't remember what's happened. The moment is tucked
away in some hidden corner of her brain. When her eyes
open, she generally finds herself under the gaze of a series of
strangers trying to help by fanning her or offering her water
and tissues. These strangers tend to try to help her piece together the time lost from her memory. You leaned against
the wall, you held your head, you vomited, you sat down on
the ground, you closed your eyes, you collapsed. A chorus of
voices offering up details of the blackout, enough for her to
partially recover the scrap of life hidden in a parenthesis in
her brain. It upsets my mother not to be able to remember
what happened in these spatiotemporal lapses. Falling down
in the middle of the street, collapsing in her seat on the bus
or in line at the supermarket—these things are less troubling
than the lost minutes of lucidity. The black holes that lurk in
her everyday memories bother her more than the bruises she
collects each time she faints.

I understand my mother. I have a theory that we're made

up of these everyday memories. It's not an original idea, but I believe it. The way we wake up, what we have for breakfast, a walk down the street, an unexpected downpour, some annoyance, a surprise in the middle of the day, a story in the paper, a phone call, a song on the radio, the preparation of a meal, the smell from the pot, a complaint filed, a scream heard. Each day and each night lived, year after year, with its full complement of activity and inactivity, upheavals and routines—continuous storing of all this is what translates into personal history. Our archive of memories is the closest thing we have to a record of identity. It's the only clue to ourselves, the only way to figure ourselves out. I guess that's why we're asked to claim it on the therapist's couch. Sorting through childhood, adolescence, youth; declaring step by step what we've lived. Because all of it—everything collected in the kaleidoscope of our hypothalamus—speaks for us. Describes and reveals us. Disjointed fragments, a pile of mirror shards, a heap of the past. The accumulation is what we're made of.

I understand my mother. Losing a memory is like losing a hand, an ear, one's very navel.

▶

On the hospital room monitor I can see my mother's brain activity. She is lying on a bed, her head sprouting electrodes and her eyes tightly closed. A series of stimuli administered by the doctor triggers electric charges in her brain. A network of hundreds of thousands of neurons interwoven with millions of axons and dendrites exchanging messages via a connective system of multiple transmitters: that's presumably what I'm seeing translated on the screen. The complexity of what's going on in there when my mother inhales, exhales, or is illuminated by the soft flickering of a light on her eyelids is indescribable. And when the doctor suggests a simple relaxation exercise, like thinking of a happy moment in her life, her brain really puts on a show. As my mother conjures some unspoken memory, a group of neurons lights up. In his office, the doctor showed us images of active neurons. Though the picture on the monitor doesn't translate those electric sparks the same way, what I see looks like a starscape. An imaginary chorus of stars twinkling softly in my mother's brain, soothing her, steadying her nerves during this test. A network stitching together familiar and comforting sensory details, I guess. Smells, tastes, colors, textures, temperatures, emotions. A neuronal circuit like the most complex stellar tapestry. In my mother's brain, groups of stars constellate in the name of the fond memory lighting them up.

▶

The last time I saw a constellation with any clarity was years ago, up north, far from the polluted skies of Santiago. I spotted Ursa Minor, Orion, the Three Marys, and the Southern Cross, which as a child I was told pointed the way home. I summon the memory and I think about the spectacle surely being staged inside my head.

A moonless night. The cold of the Atacama Desert creeping up the sleeves of my jacket. Some drowsiness, pent-up fatigue. Soreness in my neck from long minutes of gazing skyward. An astronomer indicating different constellations with a laser pointer, explaining to a group of tourists and me that all those distant lights we see shining above our heads come from the past. Depending how far away they are, we might be talking about billions of years. The glow from stars that may be dead or gone. Reports of their death have yet to reach us and what we see is the glimmer of a life possibly extinguished without our knowing it. Shafts of light freezing the past in our gaze, like family snapshots in a photograph album or the kaleidoscopic patterns of our own memory.

As we stare open-mouthed at the firmament, immersed in our genuinely Paleolithic ritual, I remember a crazy theory my mother came up with when I was little. I think it was at our Barrancas house in the port city of San Antonio, near the sea, another place you could see the stars. Sitting on the patio, smoking a cigarette on a summer night, my mother said that way up there in the night sky little people were trying to send messages with mirrors. A kind of luminous Morse code, relayed in flashes. I can't remember why she said it. She probably came up with it in response to some question of mine. What I do remember is that I assumed the messages sent by those little people in the sky were to say hello and assure us they were there, despite the distance and the darkness. *Hello, here we are, the little people, don't forget us.* They never stopped signaling. We couldn't see them during the day, but they were always there. Whether or not we looked up, whether or not we were inside our houses in the city, under a blanket of pollution, blinded by neon lights and billboards, oblivious of what was happening above, the little people's signals were there and would be there every night of our lives, flashing for us. Lights from the past making a home in our present, lighting up the fearsome darkness like a beacon.

Crazy as her old theory might be, a vague sense of peace came over me when I remembered it on that cold night in the desert. Like a whisper, like the soft voices of grandmothers singing us to sleep, like the memory my mother summoned in the examination room to try to soothe herself. A buried urge for a return to the womb satisfied by that nocturnal scene. The strange sense of an enduring, mysterious, protective reality, confirmed by each of those orbs speaking to me with their light from another time. *Hello, here we are, don't forget us.*

▶

In a life I never had I was a brave cosmonaut and I navigated the stars I'd always watched curiously. In this life I plunged into strange galaxies, witnessed the explosion of supernovas, escaped from black holes, and crossed entire nebulae; I was surprised by the dance of comets, the streaking passage of tens of meteorites, the presence of white dwarfs and red giants. I saw hundreds of stars as yet unnamed twinkling around me; I yearned to hear their dead voices, heed their cries for help. And from each stage of this voyage I never made, I sent postcards of the starscapes that sprang to life in my mind when I saw a memory of my mother's on the monitor in that hospital room.

As I learned in the desert, light from the past illuminates our present. My mother summons a happy scene from her life and the neuronal mechanism is a present act that reverberates electrically in the shape of a constellation. My mother revives a scene from her past, and the brain process is a present act as complex as the vast fabric of the cosmos that knits itself mysteriously over our heads, enthralling and confounding us. Wrong though I may be, I like to think that the human brain—the organ used by women and men for centuries to observe the universe and try to understand it—must be one of the most complex systems in the universe.

▶

Inside my mother's brain, stars constellate under the name of the fond memory that lights them up.

But what memory is it?

What piece of her broken mirror are we talking about?

▶

Epilepsy, that's what's causing my mother's disconnects. After this test and the many others that preceded it, the neurologist gives us his final diagnosis. The fits are triggered by an excess of electrical activity in one of her neuronal circuits. I imagine something like an energy drain, a cerebral short, a momentary blackout, a halt in transmissions for the duration of one of her episodes. Then brain activity is restored and my mother starts working again. Same as a house when the main fuse blows and everything stops. Clocks, televisions, radios, refrigerators, the internet, a world on pause, still and silent until someone flips the right switch and the house alarm goes off and the system resets and everything starts up again. As if the short circuit hadn't frozen anything. As if a moment of life—a hand, an ear, even the navel—hadn't been lost in that space-time parenthesis.

We exit the neurologist's office and I look at my mother with new eyes. Now I know that she's carrying the whole cosmos on her shoulders. I tell her what I saw on the doctor's screen. I tell her how much her brain looks like the night sky. I tell her about the electrical patterns of her neurons, the glow of her memory, the constellation that lit up the moment she summoned it, the luminescent reflection of her own past. I ask which happy scene it was that I saw twinkling on the monitor in the doctor's office and she smiles and says she was remembering the moment I was born.

▶

My life's first scene is a constellation in my mother's brain.
(Hello, we're the little people.)
The ground zero of my past shining in her head.
(Here we are, don't forget us.)
The Southern Cross showing me the way home.

www.constelaciondeloscaidos.cl

A few months ago I was invited to sign a petition addressed to the International Astronomical Union. My signature, along with those of anyone else interested, would endorse the creation of a new constellation in the sky. The explanation included a link to an Amnesty International web page, where more information could be found: www.constelaciondelos caidos.cl.[1]

Atacama Desert, Chile. The best place in the world for stargazing. The same place where, forty-five years ago, twenty-six Chileans were executed by the Caravan of Death. This is what I read on the screen as a video begins to play. How do we make sure the twenty-six are never forgotten? How do we make sure such things never happen again? Now images appear before my eyes of the desert, the night sky, an aerial shot of the city of Calama, where the victims were from. Next comes an array of black-and-white photographs of their faces. We want

---

1. The website www.constelaciondeloscaidos.cl no longer exists. It's a dead star whose light has yet to reach us. Some traces of this project can be found here: https://amnistia.cl/noticia/chile-dia-internacional-de -apoyo-a-las-victimas-de-la-tortura-constelacion-de-los-caidos/

to rename twenty-six stars and give them the names of the victims, I read on the screen. We plan to create the first memorial in the universe where stories and lives can be told in the stars. Next we see the wives, mothers, sisters, and daughters of the victims, sifting the desert in search of some trace of the dead they were never able to bury. The same sky that one day watched them depart today remembers them forever, I read. Learn about the constellation and add your signature to send the petition to the International Astronomical Union.

Cancer

I was born when the sun was crossing the constellation of Cancer at noon on a winter day in 1971. Early that morning my mother felt a hot liquid run between her legs and assumed it was her water breaking. It was her first and only labor. She had a suitcase ready to go and she left the house carrying it and a towel to wipe herself off. A few yards away on the same street was a garage where Don Tito came early each morning to work in grease-stained overalls. My mother asked him to help her, please. Her body was telling her it was time to have the baby, but she couldn't take a bus or get to the clinic by herself. So Don Tito, with his grimy hands and ginger mustache, got into one of the junkers in the shop and set off as fast as he could with my mother, her suitcase, and her towel.

There are more than one hundred stars in the constellation of Cancer, only fifty of them visible. The brightest of all is Al Tarf, an orange giant five hundred times brighter than the sun and 290 light-years from our solar system. My brain can't fathom exactly what that means, but I understand that its light departed a long, long time ago, making an endless voyage from the past to twinkle here over my head. What I see in the night sky, when I can see anything at all, is a bright postcard of a moment that has ceased to exist but lives on in the glow. Like my mother's memory of the moment I was born.

From the time of humanity's first recorded memories, women and men have watched the infinite. In an astronomy book, I read that our Stone Age ancestors made notches in animal bones to represent the phases of the moon. The rising and setting of the sun, eclipses, lunar phases, and the position of the stars became way-finding tools, the sky a map. I read that the orientation of the stars on this map helped measure time and the length of the seasons, and guided sailors and merchants when they traveled at night, across the sea or through the desert. Seeing figures in the stars and inventing legends and stories about what they symbolized made them easier to spot and the routes simpler to follow. I read that this is how constellations were born. A bright compass in the dark.

Cancer is one of the twelve constellations of the zodiac. In the East, four millennia before the birth of Christ, the Babylonians watched the sky and realized that the sun, the moon, and the planets rotated in front of the stars, which formed a kind of backdrop. The celestial bodies seemed to follow the same path year after year, and the Babylonians divided it into segments according to the constellations traversed. This astral route is what we know as the zodiac. I read that the word comes from the Greek *zodiakos kyklos*, circle of animals, and that it originally consisted of eighteen clusters of stars in the shapes of animals. Bull, goat, lion, fish, scorpion. Then the number was reduced to twelve to coincide with the lunar months. In this arrangement, Cancer is the constellation of the crab.

According to Greek mythology, Hera, wife of Zeus, flew

into a rage when she learned that the god had consorted with a mortal. Determined to get revenge, she unleashed her wrath on the son born of this dalliance. From the moment Hercules was born, he endured Hera's many schemes, the cruelest being the potion she gave him to drink in a cup of wine to drive him mad. Raving and enraged, Hercules mistook his wife and children for enemies and killed them. As punishment, he had to undergo twelve terrible labors. One was the killing of the Lernean Hydra, a sea serpent that lived in a swamp. It had nine heads, and each time Hercules cut one off, two or three grew in its place. To thwart the hero, Hera sent a crab to snap at his feet. Finally, Hercules seared the Hydra's neck stumps to keep the heads from growing back and pinned the crab under his foot to make it leave him alone. The hero emerged victorious from this test, but Hera decided to move the Hydra and the crab to the sky, where to this day they appear in the form of two starry constellations.

In 1971, while the sun was reportedly transiting some part of the constellation of the crab trodden upon by Hercules— maybe a leg, maybe an antenna, maybe the middle of its shell—my mother pulled up in front of the clinic in Don Tito's car, bid a grateful farewell to her neighbor, and got out lugging her suitcase and towel, alone and nervous, ready to give birth.

▶

My life's first scene is absent from my memory. No neurons constellate in my brain when I try to conjure up the moment I was born. It's the starting point of my past, the ground zero of everything, the navel of my own story, and yet it's hidden in some corner of my hypothalamus like the minutes my mother loses when she faints. I was born forgetting, it seems. We all are. No one remembers the instant they came into the world. Freud explained it as infantile amnesia, a kind of unconscious repression of our first memories, a defense mechanism that stows them somewhere inaccessible, where they can't hurt us. Later theories postulated that this forgetting of our beginnings really has more to do with biology. Before we reach the age of three, more or less, the brain hasn't developed enough to retain conscious memories. We have no capacity for long-term storage, which is why we can't relive our first steps in the world, let alone remember the moment we were born.

And yet we also have generous storytelling to illuminate memory. Our mothers, our grandmothers, our fathers, our grandfathers—they pass the baton, filling in what we lack the capacity to remember. Like the first people who told themselves stories to remember the position of the stars—tales that brought the constellations to life and guided the tellers in the night—the memories transmitted by our ancestors ground us, giving us a place in the world and a starting point on the path we might take.

▶

She remembers getting to the front desk and writing a check to be admitted to the clinic. She remembers being taken to a room where she could change clothes and leave her suitcase and wet towel. She remembers it was cold, a gray day, and she got goose bumps when she had to put on the flimsy robe she was given. She remembers being taken to another room, where she would labor. She remembers—she tells me—that all this time she was alone. No one waiting for her outside, no one sending excited or nervous messages. She remembers—she tells me—that her mother was very upset because she had chosen to let a married man get her pregnant. She remembers her mother hardly speaking to her. She remembers thinking about her dead grandmother, commending herself to her as if her grandmother were a saint. She remembers the midwife coming in and listening to the baby's heartbeat with a stethoscope, feeling the cold metal on her skin. She remembers—she tells me—the midwife deciding unexpectedly to interrupt her labor and calling somewhat urgently for the doctor. A few minutes later my mother was in an operating room, the doctor telling her he was going to give her a cesarean section because the baby was wrapped in its umbilical cord and could suffocate. He asked whether she wanted a horizontal or vertical incision, and my mother didn't know what to reply. Whatever's best for the cesarean, she said. She tells me that after just another few minutes she could no longer feel her body, because she'd been anesthetized. The doctor announced that he was about to begin the operation. My mother remembers thinking about her grandmother again, commending herself to her again, drifting back to some night when she was a little girl and her grandmother told her a bedtime story or made her favorite meal. And deep in her memories, deep in another time, she didn't feel Doctor Vinegar—that was his name, that's how she remembers it, that's what she tells me, recalling just now the fact that he was the son of the director of the

Banco de Chile—making a vertical incision below her navel. And through that incision—now a scar—at exactly twelve o'clock on a winter's day in 1971, I appeared. That's how she remembers it, how she tells it. Daughter of a single mother and a father married to another woman. Rescued from my killer umbilical cord—the curse of Hera upon me, maybe, or some suicidal urge manifesting itself from the start—all purple and dirty, just a scrap of air in my lungs, enough for me to yell and cry, setting forth.

That's how she remembers it, how she tells it.

▶

Our body's memory is made up of infinite constellations. Some reside in the cerebral cortex, in full consciousness, but others are hidden in fathomless places. There are actually memories tattooed on our DNA in a language different from the neuronal language of the brain. Stories we carry with us in our genetic makeup without realizing it. We are the sum total of hundreds of millions of years of evolution, and the memory of that process is part of us. We apply our evolutionary memory to everything we do. No one teaches us to cry when we come into the world. It's part of our inheritance. The same goes for walking, using our eyes, feeding ourselves. Every time we lift spoon to mouth, the body digests, synthesizes enzymes, extracts energy from food, following hidden instructions filed away in a kind of library containing all the memory mustered over centuries of evolution to allow our bodies to function independently. This is how we sneeze, breathe, laugh. This is how women give birth without anyone—apparently—having taught us. This is how babies cry when they come into the world, without anyone—apparently—having shown them how. This library we carry inside us is the result of the genetic story our ancestors deposited in it. We bear in our bodies hundreds of millions of stories from the past, messages passed on through us unawares, constellations guiding us and accounting for our way of being. We are vessels of genetic memories. We are molded and designed by them from the moment we're born and let out our first cry in the world. Though we don't remember it, have no knowledge of it, never file it away in our neuronal consciousness, that cry is the first use of memory in our lives.

▶

The *Chasmagnathus granulatus* is a South American crab. Its life is simple. It spends its days digging for food and trying to evade its main enemy, the seagull. When a gull drops out of the sky and attacks, the crab burrows, flees, hides. Experiments have shown that despite its simple brain, the crab has a sophisticated memory. It remembers the exact place it was attacked and it learns to avoid the danger zone. In mammals, behavior like this engages multiple areas of the brain. But the crab does the job with just a few neurons.

I wonder whether the crab that Hera set in the sky remembers Hercules's enemy feet. I wonder whether it would fight the hero again if commanded to. Its southern crab cousin would not, because memory guides its actions. It doesn't forget what's happened to it. The southern crab never makes the mistake of returning to the danger zone.

The crab is the fourth sign of the zodiac. In astrological (not astronomical) terms, this means Cancer is the sign of anyone born while the sun is apparently passing through the constellation of the crab, putting them under its influence. I understand that, astrologically speaking, from the moment I let out that first cry on a cold winter's day, I've belonged to this group.

In astrology, the position of the stars at birth determines a person's life. Just as being born in a South American hospital in Santiago de Chile in 1971 as the daughter of a single mother and a father married to another woman constitutes a departure point on the map to be traveled, astrology tells us we have a context in the universe. According to this logic, we are not immune to the energetic influence of what goes on above us. Whether we know it or not, everything happening up there made us what we are and molds us still.

Studying the map of the heavens has been a constant in human history. The more exact our knowledge of the position and progress of the sun, moon, and stars, the better we were able to divine the season for sowing, the season for hunting, the place to head toward at night, or the time for tribes to gather. But later, once women and men discovered the planets, they watched them and tried to read meaning in the stars beyond the practical, seeking to decipher their destiny. This was how astrology began, with the theory that each of the stars must have a special energetic influence on human beings.

And so the astrologers of antiquity drew up a symbolic map of the sky to interpret these influences, with specific meanings ascribed to the sun, moon, planets, and stars. Depending on the location of the celestial bodies at the moment of a person's birth, these significations intersect in multiple and complex coordinates, charting the individual's psychic universe, reading her past, and prophesying her future. Along these axes of stellar translation, the sign of the zodiac is just one instrument of analysis.

I'm remembering that night in the Atacama Desert. I remember a moonless sky. The cold creeping up the sleeves of my jacket. Drowsiness, pent-up fatigue. Soreness in my neck from long minutes of gazing skyward. I remember the same astronomer who'd been indicating constellations with a laser pointer laughing as he talked about astrology's symbolic map. Groups of stars connected at random by someone in ancient times, he said, could exert no specific influence on anyone. Those stars in the imagined shape of a crab had nothing to do with a crab and would never channel energies connected to the equally made-up story assigned to them. The stars in each constellation were hundreds of light-years from one another; nothing joined or linked them other than the wishful thinking and imagination of some ancient dreamer. The stars don't shape our psychic universe, he said. They aren't tea leaves where the future is read. And then he invented a constellation before our eyes: the constellation of Chocolate, in the whimsical shape of a mug, and predicted that in a few minutes we would be sipping a cup of hot chocolate in the cold night.

My mother used to read the horoscope in the newspaper whenever she worked a puzzle. She did it in an entertaining way, like a TV host. So every now and then—it was always after lunch—I would hear the weekly forecast for my zodiac sign from her lips. I didn't really understand what the signs of the zodiac were or how they were related to the stars, but I liked to listen to my mother's predictions, which always prophesied good things. Surprises, presents, fun visits, hugs and kisses, which in my memory were always exactly as predicted. I have photographs of some of those forecasts-come-true. My first time looking through a microscope. My first train trip. A birthday party, the house full of friends and cousins and aunts and uncles, everybody shouting my name as I sat in a chair and blew out five candles on a ladyfinger cake with dulce de leche. All those photographs sparkle in my mind's eye, and the memories my mother foretold in her horoscope readings constellate in my brain. Now I understand, as the astronomer in the desert explained, that my mother's oracular talents and her auspicious predictions were inventions of her dreamer's mind.

I'm reading an astrology manual, looking for information on the sign of the crab. What I find tells about people moved by the moon, Earth's natural satellite and symbol of all things maternal. It says that those born under the sign of Cancer and ruled by its energies are sensitive, intuitive, imaginative beings. Fearful by nature, they are always in a state of high alert, as if menaced by a great threat. They build shells to hide in, or they bury themselves to protect their shy, vulnerable hearts. They're always looking toward the past. To them the past is as important as the present. Memory steers action, because the crab doesn't forget what has happened to it. It learns from its experiences, the manual says, and it never makes the mistake of returning to the danger zone.

Wenu Mapu

Wenu Mapu is the name the Mapuche people give to the heavens. Land up above, dwelling place of the spirits of our ancestors who once walked the earth and now protect us from on high. Journey's end for those who follow the law and the natural order of things and become sun-falcons or sun-condors.

I wonder whether my mother's little people, the ones sending messages in code with flashes of light from their mirrors, might not in fact be those sun-falcons, those sun-condors.

www.constelaciondeloscaidos.cl

Not only did I sign the petition addressed to the International Astronomical Union, I sent a message to Amnesty International, declaring my great admiration for the project and telling them I was writing a book about stars and memory and therefore I felt a secret, unexpected connection to what they were doing. If my writing could be at all useful, I was at their disposal.

Days later I received a reply. It said that in all likelihood the International Astronomical Union would not agree to change the official names of the stars, despite the large number of signatures collected. Nevertheless, a ceremony for the symbolic unveiling of the new constellation would be held in Calama with the family members of the victims. In preparation, they asked whether I would act as godmother to one of the twenty-six stars bearing the names of the victims. My duty as godmother would be to write a message for the family. Something simple, to be published online and in book form.

I accepted, of course.

Scorpio

My mother has fainted again. I answer the phone and a police officer tells me she's on the Paseo Ahumada, in the center of Santiago, recovering next to him. He says she's a little agitated but despite having fainted she feels fine and insists she can go home alone. The officer doesn't think that's a good idea, so they've agreed to call me. I get in a taxi and go to her. I see her sitting on a bench holding bottles of water and packs of tissues, talking to a couple of men in uniform. She explains that when she came to she was on the ground and a group of people were staring at her. Because she had vomited, they cleaned her up with bottled water and tissues and then they left her plenty more just in case. The officers weren't there when she fainted, but they pieced together the story my mother doesn't remember. The lady hugged the wall, held her head, vomited, sat down on the ground, closed her eyes, collapsed. They transmit the version presented by the street chorus, assorted details that help my mother retrieve the scrap of life now hidden in a parenthesis in her brain.

We go to a café to get something to drink. I want a doctor to look at her, but she refuses. She's sick of tests, she's sick of doctors and hospitals and appointments. She's sick of fainting. She's sick and tired. She just wants to drink her tea and go home. I watch her as she talks. She's pale, dark circles under her eyes, her liver-spotted hands trembling slightly. Her hair is gray, her back hunched, her green eyes now reddened and glassy. She's lost weight. I don't know how much, but her clothes are loose on her. I remember the woman who smoked a cigarette under the starry summer sky of the port city of San Antonio. The woman who read the horoscope after lunch. The woman who worked near here, just a few blocks away, at a dentist's small office, where she saw patients all day long, straining her back until it was humped the way it is now. The woman who took me to light candles before an icon of Santa Rita de Cascia, patroness of lost causes, to give thanks for secret wishes granted. The woman who one day decided to become a mother and make a space in the world for me. The woman who twinkles, hidden in some part of the person sitting across from me. The woman who is a memory constellating in my brain as I watch this other person sipping a cup of tea and giving me a tired smile. Sick and tired.

One day I'm going to faint, she says, and never get up again.

▶

Stars share a common fate. It makes no difference how strongly they resist the force of gravity, how great their initial mass is, or how many years they've shone. Every single one of them will break apart. In another astronomy book, I read that a star's life is bounded by two great collapses. The first is its birth. Gestating stars lie cradled amid masses of gas and stardust in a kind of stellar nursery or ward called a nebula. Here they wait for the temperature and density of the inner core to increase until reactions in the nucleus are triggered. These nuclear reactions are what generate energy, producing such high temperatures that the star would explode if it weren't for the force of gravity. If I understand right, the explosive force of the nuclear reaction creates pressure outward, which is contained by gravity pushing inward. Perfect equilibrium between inward and outward forces signifies the birth of a star.

I read that each star is defined by the mother nebula that cradled it before it was born. From her, it inherits the type of fuel it will use over its lifetime. The quantity it receives and the star's mass at birth shape its story, determining the moment when its inherited fuel will run out, its energy will be extinguished, and it will break apart, meeting the fate allotted to it. It's during the long stage between the two collapses that stars shine in a stable way and we can see them from Earth, in the future. Depending on how far away the star is, according to the astronomer in the desert, we might be talking billions of years in the future.

I read that the infancy of a medium-sized star like the sun can last tens of millions of years. In this first stage, the stellar dust surrounding the star at birth begins to condense under the force of gravity. Planets are formed from this condensed dust. This is how Earth came to be. And with it, life as we know it. All generations of living beings on the planet have been nourished by the energy of our star, the sun. And so I realize that women and men—every one of us—owe their existence to the birth of a star.

Once the stored fuel inherited from the mother nebula is consumed, stars proceed from adolescence to maturity and maturity to old age in rapid and violent succession. Some of them, like the sun one day perhaps, sense the approach of death and tremble as it arrives, quaking as if in fear. Then they surrender themselves to what is to come, expanding and cooling, shrinking and coming apart layer by layer. They shed parts of themselves. They fall into dust. They are dispersed in space, forming other nebulae for the nurturing of new and future stars. Tomorrow's stars, with planets like ours orbiting them. And on those planets—why not?—life arises.

Stars die and return part of their substance to the universe. Which in turn was once the substance of a star from a previous generation. Stars are made of the stellar dust of their dead. They shine in our present, glowing with the experience of the hundreds of thousands of millions of cosmic generations preceding them, a relay of light in which death is just a way station.

▶

My mother will be eighty years old this spring. One day in November 1938, when the sun—our star—was apparently passing through the constellation of Scorpio, my mother came into the world. I don't know any of the details. There are no photographs, no written accounts. I remember hearing that it was a home birth, as was the way back then, long and difficult, attended by a doctor, and that my grandmother always complained about her daughter's big head, an important and painful obstacle at the crucial moment.

That formidable head is where my mother houses all her conscious memories. A complex stellar skein spun from infancy to old age. The images are manifold, encompassing her ancestors and her descendants. We all constellate in that brain of hers and we endure, shining on, because she sustains us. At her age, the exercise of memory is hard work. Summoning images from the past takes conviction, time, and willpower. She clings to the task day after day as if to a life preserver. Like I do, she has a theory that she's made of her own memories, and she believes that if she surrenders to oblivion she will be lost, shedding layers into outer space and breaking apart. Her core energy, though diminished now, is in perfect equilibrium with the force of gravity and keeps her whole. There is still fuel in her—probably the last of it—and with it she dutifully sets aglow each of the stars constellating in her brain.

Though she's sick and tired, though she doesn't want a birthday celebration, I think she might enjoy a small gathering of family and close friends. Something I organize myself, something that won't make any work for her, a surprise. For the party, I'm collecting photographs of her life. I want to make a kind of visual history with as many snapshots as I can find. Without explaining why, I ask her to help me go through the pictures she's kept for years in an old Calpany shoebox.

The oldest are from Tunekawa, a portrait studio that once stood somewhere in the heart of Santiago. Countless people had photographs taken there, much like the one I'm looking at now. In it, my mother is very young, maybe two years old, sitting on a cushion and holding a few balls, with a big bow atop her curly hair. Her head is big. I can see my grandmother wasn't kidding. My mother is laughing, happy, playing with the balls, smiling a toothy smile.

My mother has no memory of this moment. No neurons

constellate when she tries to conjure up an image from a time before her brain had the capacity to store long-term memories. What she has instead is this photograph and the story her own mother told about it. They've left her with a sense of the moment, hidden in a parenthesis in her brain. The story and the photograph were fabricated in the past, but here they are, pulsing in my hands, vibrating in my ears, as my mother tells me about that first visit to a portrait studio.

In another picture I see her dressed in white. She has a veil over her head and a lily in her white-gloved hand. She must be about eight. Ribbons and little holy medals hang around her neck. She looks like a miniature saint, or a small Virgin Mary, even a young soul in purgatory. Behind her, floating in a mysterious celestial haze, is a blurry image of Christ. A disturbing apparition, gauzy and ephemeral. The photograph marks her first communion and she looks nervous, barely smiling. I can see why. Everything in the portrait is a little somber.

In another photograph I see my mother in my grandmother's arms. She must be about six, and they're on the patio of the Barrancas house, in the port city of San Antonio. It's her birthday, a spring afternoon in 1940-something. The family has gotten together to celebrate. A crowd of uncles, aunts, and cousins smile cheerily into the camera as if they know I'm spying on them from here in the future. Uncle Tote, Aunt Victoria, La Maria, El Cano, El Choche the Little, Uncle Beto, Don Arturo, my grandmother. They are young and beautiful, they have their whole lives ahead of them, and they seem to know it. They love this little girl very much and they've gathered to celebrate her. They will eat cake and sing "Happy Birthday" and blow out candles, all without considering—or even realizing—that at this very instant the sun, our mother star, is apparently transiting the constellation of Scorpio. Maybe it's passing over a leg. Or a segment of an antenna. Or the deadly tip of its poisonous tail.

Except for my mother, each and every one of the happy, smiling people in this picture is now dead. She is moved to see them here in the past, so very present. I examine them carefully and I'm convinced I look like all of them. I see myself in my grandmother's mouth, my great-aunt's laugh, my uncle's eyes; I recognize some of my son's features in my mother's baby face. Each face is a new glimpse into a common mirror. Distorted reflections of reflections of reflections. A whole genetic inheritance dispersed. A family lineage that began a long time ago with her grandparents, or rather the grandparents of her grandparents of her grandparents, who are mine too. And even further back. A biological relay race that everyone's been running for a long time, in which I'm just a brief stop.

A vague sense of peace comes over me as I look at this family photograph. Like a lullaby, like the soft singing of grandmothers that helps us fall sleep. Some strange awareness of a lasting, mysterious, protective reality, evident in each of these faces smiling at me from another time. Little broken mirrors, shards of glass, a scattered and shimmering puzzle in which I read: *Hello, here we are, don't forget us.*

▶

In every culture's memory, the first thing was chaos. Amid this chaos came a moment—though it wasn't actually a moment, because in this chaotic time there was no time yet—when something exploded. After this great explosion, which was the beginning of the universe, many millions of years went by during which the sun, our mother star, was cradled in a nebula, in the stellar substance of an earlier star, later to be born and to shine as it has ever since.

Women and men have always watched the sun and the rest of the stars, seeking an answer to the vast enigma of their existence. Who are we? Where are we going? Where do we come from?

I read that in the sun's early days, Earth was formed from stardust swaddling the infant star. First the planet was a red-hot globe, and as time passed it cooled and turned solid. Over several billion years, the pull of gravity caused heavy materials to be deposited in its interior while the lightest elements remained on the surface. Thus Earth's core was formed. At the same time, volcanic eruptions released vapors and gases, creating a brand-new atmosphere. When the surface temperature fell below the boiling point of water, great quantities of vapor condensed and heavy rains fell, eroding the rocks of Earth's core and making the oceans. It was in these waters, some three billion years ago, that the first living organisms appeared. These organisms evolved, producing a molecule capable of making copies of itself. From the subsequent union of a cluster of these molecules, the first cell was born. Many more millions of years later, the evolution of these cells led to the development of beings capable of exchanging genetic information among themselves—in other words, of reproducing sexually.

What comes next, like what came before, is a long sequence of chance occurrences. A chaotic and haphazard path, crisscrossed with unpredictable logic. Countless actions and reactions tracing a route full of random mutations in our hereditary material. The death of a huge number of life-forms ill adapted to the environment. The emergence of others that survived. All the result of fortuitous events that unexpectedly gave rise to the human type we belong to. If any tiny piece of that crazy game had gone down a different path, another kind of being might be writing or reading this book.

The first vertebrates and the first fish appeared. Plants, which had once lived exclusively in the oceans, began to colonize dry land. The first insects evolved and their descendants became the first animals to inhabit Earth. Winged insects arose at the same time as amphibians, creatures capable of surviving on earth as well as in the water. The first trees appeared, and the first reptiles. The dinosaurs evolved. Mammals emerged, and then the first birds. The first flowers sprang up. The dinosaurs went extinct. The first cetaceans arrived, ancestors of dolphins and whales, and the same period saw the birth of primates. The Ice Age was about to begin when a very evolved primate—capable and sociable, even agreeable, with a crucially bigger brain—staked her claim. She came down from the trees, stood on two legs, and walked for years until she evolved into the kind of human being we are today.

This whole mediocre, imprecise summary I've given in little more than a page can hardly convey the amount of time and work it took the human race to establish itself and become what it is today. So many millions and millions of years that my brain can't fully fathom how much life and death there was in between. The incredible process of learning to stand upright, think, talk, tame fire, invent tools, organize ourselves, sow the earth, domesticate livestock, build houses, found cities, forge metal, gaze up at the heavens, and wonder: Who are we? Where are going? Where do we come from?

Stardust, gas, hydrogen, fire, atoms from stars, molecules, microorganisms, bacteria, cells, algae, plants, oxygen, insects, amphibians, reptiles, invertebrates, vertebrates, explosions, gases, volcanos, rain, floods, climate change, meteorites, mammals, primates, centuries and centuries of evolution, and all I can make of it is that we are the children of chance. We move through time propelled by the energy of the great explosion that made us, swept along in the undertow of the great initial stampede. And thus we get up to make breakfast each morning, oblivious of how our organs, our genes, our race evolved despite us. A vital force from the past secretly sets us in motion, transforming us into something beyond our comprehension. We are inheritors of an enigmatic process, of a timeless time, of a logic silently proceeding along its unforeseeable path.

All this experience of all this vast past is stored in our bodies' memories, driving us forward. We carry it with us as our legacy and we draw on it daily, though we may not acknowledge it or even realize it. It's what wakes us up each morning. It's what sets us in motion: crying, yelling, eating, defending ourselves when attacked. It's what makes us gaze at the sky, quietly convinced that a reflection of us is shining up there. Thanks to it, we live with an inkling that a part of ourselves was left scattered in the infinite after that first explosion back in the beginning. And up there, in that piece of ourselves shining in the darkness, is the answer to each and every enigma we've borne with us for centuries. Who are we? Where are we going? Where do we come from?

▶

One July night eighteen years ago I had a dream. In it I returned to the house where I was born. I rang the bell and from the other side of the pebbled glass I heard the steps of people hurrying to let me in. When the door opened I saw my grandmother. And her sister, Aunt Victoria. And her brother, Uncle Tote. Everyone from my mother's birthday photograph was there, smiling, happy to see me. They welcomed me in and walked with me down the long hallway to the dining room, where they had laid a table with deviled eggs, sweet bread, sopaipillas, even a cake, I think. And to drink there was tea served in my grandmother's old flowered teacups, which I wouldn't have remembered if I hadn't seen them on the table in my dream.

My grandmother had died a few years before this dream. Soon after that, Uncle Tote was gone, and months later, Aunt Victoria. In the dream I knew this and I stood before them perplexed, as anyone would be when faced with three ghosts. I was perfectly aware that I was dreaming, and I understood that this reunion was a gift to be cherished because it wouldn't come again. The dream was so clear. I could hardly take in everything I was experiencing. I wondered where in my brain I could have filed away so many details. The scent of my grandmother's cologne. The shape of my uncle's glasses. The clips my aunt wore in her hair. It was so powerful, my mind dredging up memories I had no idea were still in me.

But the voices were what surprised me most. Pure, pristine, sounding in my ears in a way I could never reproduce in my head. Their genuine timbre, vibrating as in life with that uniquely identifying character of a person's voice. It was as if they were really there, not a mere fantasy of my unconscious, not a product of the neurons wringing my memories, no mere projected constellation of names, faces, voices.

What's the difference between dream and memory? Is there some boundary separating the two? Some territory they don't share? Basted together with threads of air, sliced apart with the delicate edge of a scalpel. Volatile, muddled, capricious, so easy to confuse and entangle that maybe there's no point drawing distinctions. It was their voices. They were talking about things I didn't quite understand, but I went with the flow because they were old, and toward the end I always went with the flow, ignoring their bewildering, daffy reasoning. I would smile and nod, just as I was doing now on this visit, when everything was so disturbingly familiar.

At some point my grandmother fetched a bottle of red wine, saying we had to celebrate. I didn't quite understand what we were celebrating, but following the usual script, I said, Yes, let's raise a glass, suspecting they might have mistakenly thought it was my birthday. And then a couple of gifts emerged, wrapped in plastic bags. One was from my grandmother and the other from my aunt. They watched me closely as I pulled out two necklaces made of little white shells with fish bones dangling from them. They were identical. Horrible and identical. I pretended to like them, as I always did with their gifts. Nothing could be better, I said, than these two necklaces with fish bones dangling from them.

The three of them clapped happily as I discovered another necklace in my grandmother's gift bag. This one was small and fine, so delicate and fragile that it really did interest me. My grandmother helped me fasten it around my neck and I left it on. I wanted to keep it. It wasn't just an act. I liked it. Then the three of them hugged me and congratulated me. And I felt so happy to be there with them again, swaddled in their old-person smells, their old-person arms, their old-person love, like a budding star in its mother nebula, surrounded by the gases and stardust of earlier generations. And I thanked them for the gifts, the wine, the food, the tea, and this whole celebration they had planned for me, though I didn't understand it. I really didn't understand it.

I woke up from that dream and I wrote it down right away. I don't usually write down my dreams, but for some reason this time I did. Part of what I wrote is what I've just set down here. At the end of the final paragraph, I wrote: I still don't know what we were celebrating, but some part of me was expecting it. And after that I noted the hour, the date, and the place I had the dream.

Two weeks later I learned I was pregnant. If all went well, as the sun was apparently crossing the constellation of Aries next fall, I would become a mother.

www.constelaciondeloscaidos.cl

Star HD89353: Mario Argüelles Toro. Star HD90972: Carlos Berger Guralnik. Star HD85859: Haroldo Cabrera Abarzúa. Star HD70523: Carlos Escobedo Caris. Star HD70442: Jerónimo Carpanchi Choque. Star HD70761: Bernardino Cayo Cayo. Star HD72310: Luis Alberto Gahona Ochoa. Star HD73495: Daniel Garrido Muñoz. Star HD73752: Luis Alberto Hernández Neira. Star HD73840: Manuel Hidalgo Rivas. Star HD74745: Jorge Rolando Hoyos Salazar. Star HD75605: Domingo Mamani López. Star HD78541: David Miranda Luna. Star HD78702: Hernán Moreno Villarroel. Star HD80479: Rosario Muñoz Castillo. Star HD80586: Milton Muñoz Muñoz. Star HD82205: Víctor Ortega Cuevas. Star HD82232: Rafael Pineda Ibacache. Star HD72908: Carlos Piñero Lucero. Star HD82734: Sergio Ramírez Espinoza. Star HD83380: Fernando Ramírez Sánchez. Star HD83754: Alejandro Rodríguez. Star HD84117: Roberto Rojas Alcayaga. Star HD84367: Jorge Yueng Rojas. Star HD866267: José Saavedra González. Star HD80479: Luis Moreno Villarroel.

▶

When I was very little and I asked my mother about the stars, she responded with a crazy theory. Up in the night sky, she said, there were little people who were trying to talk to us with mirrors. In a kind of Morse code with flashes of light conveying messages. For a long time I believed her and I assumed that the messages sent by the little people in the sky were to say hello and remind us of their presence despite the distance and the darkness. Hello, here we are, we're the little people, don't forget us. They never stopped signaling, even when we couldn't see them during the day. They were always there. It didn't matter whether we were looking up or not, whether we were inside our houses in the city, under a blanket of pollution, dazzled by neon lights and billboards, oblivious of what was happening above our heads; the messages were there and would always be there every night of our lives, lighting up for us.

Now I know that Mario Argüelles Toro is there, in Star HD89353, signing to us, trying to make contact with his shards of broken mirror. The light of his past settled in our present, shining like a beacon in the fearsome darkness.

My warmest wishes to his family and especially to Violeta Berríos, his widow.

Nona Fernández Silanes

▶

I'm the godmother of a star to be named Mario.

I'm the nebula where preparations are being made for the long-awaited birth.

Pisces

The last constellation in the zodiac is Pisces and it is made up of approximately 150 stars. It looks—with a little imagination—like two fish tied with a rope at their tails. The fish are heading in opposite directions, but they can't be separated. They are joined by the rope.

According to Greek mythology, the fish are Aphrodite and her son Eros. Both were fleeing the dreaded Typhon, a giant so huge that when he stood on Earth his enormous head reached the stars. The goddess Gaia, angry with Aphrodite because she'd had an affair with Ares, sent Typhon after Aphrodite and her love child. Aphrodite turned herself into a fish and leaped into the water with her son to hide from the terrifying monster. To keep the current from separating them, Aphrodite tied one end of a rope to Eros and the other to herself. No matter what happened beneath the dark waters, they would be together. Surprised by this heroic feat, the god Zeus decided to lift them up into the sky, where they remain to this day in the shape of fish.

Mario Argüelles Toro, Star HD89353, my godson, was born on March 5, 1939, when the sun was apparently transiting the constellation of Pisces. Maybe it was crossing the aqueous eye of one of the fish. Or an end of the umbilical cord joining them, or a silvery scale on one of the fishes' tails. I don't know what Mario's arrival in the world was like. There's no one to ask, and no way to find out. I imagine it was in Calama, but I could be wrong. I search for information about Mario on the internet but don't find much. His name turns up on a web page alongside the names of the other twenty-five executed by the Caravan of Death in Calama. An attempt at a sketch of him in three lines. Thirty-four years old, salesman and taxi driver, socialist leader. Arrested on September 26, 1973, days after the military coup, and sentenced by the October 16 war council to three years of banishment south of the thirty-eighth parallel. On the date of his arbitrary execution he sat in jail waiting to be transferred to the place he would serve his sentence. In some parallel life, if the Caravan of Death hadn't come through Calama, Mario would have lived for three years in the south, far from his family. Then, in the best-case scenario, his banishment would have been lifted and he would have gone home to Violeta, his wife, and I wouldn't be here, searching for him on the internet and trying to write about him.

When I sent my message of support to Mario's family, I also wrote again to Amnesty International. In my email I asked if I could attend the inauguration of the new constellation. If I was going to be a godmother to a star, I wanted to be present at its birth. I soon got a reply saying I was welcome to come, and if I wanted to I could join the group organizing the event. I liked that idea and a few days later I was working with the team. There were meetings, duties assigned, emails. Messages to the families were collected, a book of tributes designed, a wall hanging embroidered with a scene of a starry night in the desert, a folk singer hired, the script for the event composed, and a technical crew, lights, audio equipment, chairs, and buses to transport the families procured. For weeks I was wrapped up in this group creation activity, organizing materials, working hard to prepare for the ceremony, part of the nebula collectively cradling a star about to be born. Or, in this case, twenty-six stars.

Mario Argüelles lived at 2301 Calle Hurtado de Mendoza in the city of Calama. Since we were making the trip north, I wanted to meet his widow and now I'm here at his front door. In a parallel life that Mario never lived, maybe he would come out, shushing the dogs barking at me through the bars. He would be seventy-nine, looking different than he did in the picture I found on the internet. Definitely grayer, stooped, a little heavier or thinner, maybe a pair of glasses perched on his nose that he didn't live long enough to need. In this parallel life Mario never had, maybe he would be surprised to see me. Excuse me, miss, how can I help you? he might ask uncertainly, gazing in bafflement at my ghostly presence, not suspecting that in this other life we're leading, he doesn't exist anymore and all I can do is imagine him on the page, in this book.

Violeta, his widow, comes out instead. The dogs dance around her, wagging their tails, obeying her when she tells them to quiet down. There are three of them: Luis Miguel, Martina, and La Cata. I told her I'd be coming so she's expecting me and welcomes me in, ready to talk about Mario. She's used to telling his story, talking about the Caravan, her life searching in the desert. We sit down in the living room.

The canine committee is sprawled on a big bed beside us. Young Martina is the most active. She weaves between our legs and walks in and out of Violeta's bedroom, the same room Violeta once shared with Mario. The other two dogs are listless. I don't know what's wrong with them, but they move slowly and hardly bark. On the walls, the coffee table, and a bookcase next to me are photographs of a little dog, Shakira. There must be six or seven of her in different scenes and settings. Shakira in the desert with a red kerchief around her neck, like the women of the Association of Relatives of the Detained–Disappeared of Calama. Shakira with red bows on her ears. Shakira in a chair at home with Violeta; on the streets of Calama; at the memorial site where the star ceremony will be held a few hours from now. Violeta tells me Shakira was her baby girl—that's what she calls her. They were together for eleven years, and when the dog died she sank into depression, a sadness and emptiness she had never felt before. She was sick to her stomach and she lost weight. Violeta is thin and when she talks about her dog she gets very emotional. She says Shakira is irreplaceable. Martina, Luis Miguel, La Cata—no dog can take her place. She misses her terribly and nothing will ever be the same without her.

Behind us, in a less prominent corner of the house, I spot a photograph of Mario. It's a blowup, low resolution, but better than the picture I saw on the internet. I go over and examine it carefully. He's wearing a three-piece suit and tie and he's sitting in some indeterminate place, looking to one side with a modest smile. He's dark-skinned, his thick hair combed back, a nice face. It's a good picture. Violeta says Mario was thirty-three when it was taken.

Hello, Mario, I think. I'm your godmother, here I am, I've been waiting to meet you.

Violeta remembers it was an April night when she saw him for the first time, at the corner of Calle Exposición and Alameda in Santiago. She remembers she was waiting for the bus and he asked her for the time. She remembers looking at him and thinking he was handsome. He was tall and dark, and she had already been secretly watching him at the bus stop. She remembers that when she got on the bus he did too. She remembers that when she sat down, he sat next to her. She remembers that they talked during the trip and that when she got off he asked whether he could walk with her. She remembers—she tells me—that her stepmother was waiting for her at the front door because it was such a quiet block. She remembers—she tells me—that Mario walked with her and when they got to her house she had to introduce him even though she didn't know his name yet. She remembers that her stepmother asked him in and they ended up sitting around the table talking and eating. She drank milk, she says, two glasses of milk. She remembers—she tells me—that the next day Mario had to go back north on the train. He was a salesman, he made trips to Santiago to buy merchandise to sell later to kiosks in Calama. Bars of soap, thread, pencils. But Mario didn't go back. And the next day he came to see her. And again he was asked in and again they sat at the table and drank more milk and talked and this—Violeta remembers, Violeta tells me—is how the courting began.

Remember. *Recordar.* From the Latin *recordari*: the prefix *re*, for repetition, plus *cordis*, which means heart. Etymologically speaking, then, *recordar* means to channel back through the

47

heart. So if a constellation of neurons lights up in some part of our brain each time we remember, then presumably the brain and the heart are closely linked, like two fish tied to each other by the tail.

Years later, Violeta and Mario moved to Calama. They lived in the same house where we're sitting now, remembering. They slept in the same bedroom I can see from here, Martina walking in and out as if it were hers. They cooked in that kitchen, ate at that table, and more than forty-five years ago, under this very roof, they planned and tried so hard to have a daughter or a son whose future pictures—in the parallel life Mario never lived—would sit next to or replace those photographs of Shakira. Then one day in September 1973, while Violeta was in Santiago at a fertility clinic appointment, the military coup arrived and Mario was detained in Calama. Then came the torture house, the jail where Violeta brought him food and clean clothes, his sentence of banishment, and the unexpected visit of the Caravan of Death.

When must Violeta and Mario have tied the rope between them so as not to get lost? When must they have become aware of the threat of monstrous Typhon? After October 19, 1973, when Mario was executed in the desert, Violeta searched tirelessly for him. Day after day for twenty years she went out early into the pampa, combing the desert with her hands, seeking a bone belonging to Mario. A splinter of his body, a scrap of his clothes, something so she could bury him and bid him farewell. She remembers—she tells me—that time was sealed in the search. She was frozen in a space-time parenthesis. You know what? she says. I didn't notice when I turned forty. Or fifty. Or sixty. Or seventy. I was on pause, somehow. All I did was look for him. Life went on around me, but I was never aware of it.

Now Violeta is eighty, just as my mother will be in a few days. Violeta looks at me and tells me she's tired. Sick and tired. A psychiatrist has recommended she let Mario go, leave the past behind; her body and mind can no longer sustain the umbilical cord that joins them. It's too much for her, too heavy, in these dark waters that she's had to throw herself into. Better to get her head above water and come up for air. Breathe. Gaze at the night sky and imagine Mario floating free, far away, like a kite whose string has been cut. There he is: his kind face, his modest smile, his combed-back hair. He swims from star to star, pausing every so often to glance at Earth and send signals with a shard of his broken mirror. From up above he beams a message in code. Probably: *Hello, Violeta. Here I am. Don't forget me.*

▶

Many would ask: Why do you want bones?
I want them, Violeta says.

▶

Thirteen kilometers southeast of Calama, on the road to San Pedro de Atacama, stands the memorial to the twenty-six who were executed. In the process of organizing the star ceremony, we've been looking at pictures of the place. Twenty-six wooden pillars stand on a semicircular concrete base, facing a big iron cross. On each pillar is a plaque with the name of a victim and his birth and death dates. The monument was inaugurated on October 19, 2004, thirty-one years after the crimes.

We're heading along the highway toward the memorial site in a van loaded with everything we'll need for the ceremony. It's early. The idea is to have enough time to get organized before we welcome the buses bringing the families and everyone else who wants to participate. We've been driving for quite a while, and the thirteen kilometers that remain between us and our destination seem to have stretched out. Maybe the monotonous desert landscape is playing tricks on us and we've stayed on this road too long. We ask the driver whether we're going the way we're supposed to, whether he's sure we've taken the right road, and he says yes. He's a local, born and raised in Calama. He has experience on these roads. He knows them; he's driven them all. He makes a confident right turn, leaving the highway behind and heading into the desert.

The past lives in this landscape. Traces of what happened appear like clues in a game we can't help playing. What happened here? How? When? Why? The desert climate presents archaeologists with the prospect of an infinite universe to explore, like outer space itself. The Atacama is the driest desert in the world. Average humidity in its interior is 18 percent, which means all organic decomposition processes are slowed. The abundance of salt helps mummify bodies, keeping objects intact for a long time, immune to the passage of the years.

It's the dry climate that makes the desert sky thin and transparent. And it's this transparency that makes the Atacama the perfect place to observe the firmament. The elevation, the absence of clouds, the nearly nonexistent humidity, and the remoteness of artificial light and radio pollution mean pristine visibility in the night skies. Which is why there are more than a dozen observatories here. Forty percent of the world's astronomical observation takes place in Chile. And the development of new sites over the next few decades means that nearly 70 percent of the world's sky gazing will be concentrated in the north of the country. If we recall that everything we see in the sky is part of our past, we must accept the idea that the Atacama Desert is the planet's most important portal for time travel. Whether we're watching the ground or the sky, the Atacama is the port of departure for a cruise to what's already happened: square one of the game we can't help playing. What happened here? How? When? Why? Neuronal magic lighting up and constellating the memories of the planet.

Mirages rise off the road in the afternoon heat. Fantasy of a liquid scene that isn't there, though it's visible from the distance. An optical illusion, apprehension of a near and watery future that recedes the closer our van gets. We're crossing an enigmatic land, a mirror zone reflecting scenes launched from the edge of time. If I close my eyes, I hear the sound of a helicopter flying over the desert, and if I concentrate even harder, I catch the strains of a military band playing its terrible music. Maybe I'll hear the voices of the recent arrivals disembarking, curved knives hanging from their belts, adorning their unsettling combat uniforms. Maybe I'll hear their heavy boots on the concrete. A welcome address, the requisite salutes. The engines of trucks driving along the streets of Calama, the nervous shuffle of detainees exiting the jail, their rapid breathing, a question gone unanswered. Maybe then I'll hear the transfer to the desert. The screams, the shots, the rattle of the machine guns. These are sounds from long ago. I'd rather not hear them, but they're still here, lurking in these empty hills, and there's no escaping them.

I know that the detainees were led out of the jail in groups. I know they were put into trucks and vans and taken into the Topáter hills, a few kilometers from Calama. I know they were unloaded there, and gathered to be executed. As if facing an enemy army, not a group of bound, unarmed men, the soldiers used their curved knives and their machine guns. A single shot wasn't enough; they had to unsheathe their sharp blades. The bodies fell one by one onto the arid ground. And one by one they were collected and piled into a couple of trucks. The scene is part of a mirage still being projected somewhere in these hills.

What did they think before they died? What did they feel? What did they hear? We can ask ourselves. We can try to imagine. We can baptize stars with their names, erect monuments, light candles, bring flowers, turn them into souvenirs, and even print cards or pins stamped with their faces, but we'll never know anything about that private final moment. Not even the desert has an answer. Repelled by its alien landscape and powerless to express myself in a language beyond my command, I try to focus on Mario. His dark, pleasant face, the modest smile in the photograph on the wall of his house. It's hard for me to imagine him that October day, facing his executioners.

In 1990, after years of searching tirelessly for a single bone, Violeta and the other women combing the desert found something. Splinters, fragments, scattered bits like the memories stored haphazardly somewhere in their hypothalamuses. Tiny pieces of beings who were no longer there, but who existed in some part of them, tied by the tails to their bodies, to their brains, where they were filed away.

Excavation sites were set up, where the women found the bones and archaeologists and experts were able to reconstruct events. They probed the space-time parenthesis in the pampa where this one particular moment was hidden. What happened here? How? When? Why? The traces of the past were fresh, as if the trucks had driven through yesterday, as if the machinery for digging up bodies had been brought in just a few hours before. Here is where the victims were brought from the execution site, here is where they were buried and hidden that very same day. And it was from here, years later, that they were taken to be hidden forever, so no one could ever find them. Buried and then unburied. Hidden and then re-hidden. Disappeared and disappeared again.

The military used bulldozers to dig up the bodies from the secret original grave. As they were being transferred to new trucks there was some jostling of skulls and limbs, and the remains scattered by the machines were found by the families in 1990. The site of that first pit is where the memorial we're heading to was built. The plaques with the names of the twenty-six victims are there because it's the closest thing to a grave their families have. It's the only place in which they know for sure their family members were once buried.

The driver tells us not to worry, we're very near the memorial site. And yet no matter how far we go, there's no glimpse of anything like what we've seen in the photographs. No iron cross, no group of twenty-six pillars. One of us was here yesterday. She came to check things out and she keeps telling the driver that everything she sees looks a little different.

The desert is the same everywhere and it's hard to spot landmarks, but there aren't any of the wind turbines she's sure she saw along the road, she says, unless they were also a mirage. The driver reassures her and brakes. This is the place, he says. We're here. We look out the window, and next to a lonely, dilapidated cross, a hand-painted wooden sign reads: Cementerio de Perritos.

A charged silence descends on the van. We exchange glances of puzzled complicity. I feel like laughing or crying, but I restrain myself. I don't know what bizarre connection this man has made between a memorial to political victims and a doggie cemetery, but clearly he's made a mistake. We explain to him again where we're going and he stares at us in bewilderment. He doesn't know what we're talking about. He never did. He's never heard of the memorial. He's probably never even heard of the Caravan of Death, or Mario, or the other twenty-five victims.

▶

A piece of Mario's jawbone was found.

I don't want it, said Violeta.

They took him away from me whole and I want him back whole.

▶

At the center of the Campo de' Fiori, in the heart of Rome, stands a statue of Giordano Bruno, philosopher, theologist, astronomer, physicist, and poet. It's a popular spot, with a market and restaurants where tourists eat pasta and photograph themselves next to the towering figure. Born in 1548 in the town of Nola, near Naples, as the sun—our mother star—was apparently crossing one of the constellations that he himself observed, Giordano was one of the first to maintain that our planet circled the sun and that the turning of the heavens was only an illusion caused by the rotation of Earth on its axis. Following the theories of Copernicus, Giordano declared that our planet was not the center of the universe and that each and every one of the bright stars seen in the night sky was another sun, around which other planets probably orbited, planets that probably harbored other lives, because the universe was infinite. So it was with the human intellect, he believed. The mind was the mirror of the cosmos and its possibilities were limitless. The art of memory permitted women and men to store absolutely everything, and with this invaluable tool they became bearers of absolute knowledge, transcending humanity itself. These and other revolutionary ideas challenged the spiritual and theological beliefs of the era, leading him to be persecuted, arrested, tortured, tried, and sentenced by the Roman Inquisition. Judged an irreverent, inveterate, and unrepentant heretic, Giordano Bruno died one February day in 1600 in the Campo de' Fiori, burned at the stake.

There's a waxing moon and we won't be able to see the stars as well as we'd like to. Also, at this time of year the twenty-six stars of particular interest to us won't appear until later, around four in the morning. An astronomer is with us and he has prepared a speech to explain this to the families. They've come in two buses and a few vans to a place in a forgotten corner of the desert, altogether unlike the monument to Giordano.

At last we're at the memorial site. After roaming the desert we've managed to arrive and get everything ready. We've set up lights and a sound system, a table of food to share, coffee to keep us warm in the cold of night. The families have brought photographs of the twenty-six victims, which we've arranged in a circle with flowers and lit candles. A folk singer has performed a song composed for the occasion, and a quilters' collective has made a delicately stitched wall hanging showing the constellation of the fallen in the desert sky. There are speeches. There are heartfelt words. There are video messages from many people wanting to send their regards. Artists, politicians, writers, a whole spectrum of citizens. It's all very poignant, and my spy-gaze softens when I see how important this moment has become for the families, especially the older women like Violeta, who sits next to her niece in the front row.

Now it's the astronomer's turn. He's a young man, probably not yet forty. I don't know him, but I've had my eye on him for a while, because despite the desert cold he's come in shorts and a light jacket. The rest of us are wearing parkas and blankets to protect us from the icy wind, but he seems immune to the temperature. He approaches the microphone and explains that the stars we're here to baptize will come out later, long after midnight, and he won't be able to show them to us with his laser pointer, as we'd probably expected, but instead he's brought photographs of each one so we can get a look at them. They're on cards printed with a special Amnesty International graphic incorporating the name of each star and each victim. Souvenirs we'll be able to take home and put on the refrigerator, use as a bookmark, or simply look at whenever we want to remember this day. In addition to the photographs, he says, he wants to share a story about a thinker from long ago, a man by the name of Giordano Bruno.

The astronomer in shorts tells us that young Giordano got his start as a Dominican monk. He took orders committing him to a search for knowledge, became a theologian. The astronomer tells us that Giordano's thinking soon began to discomfit the religious authorities, so he shed his robes. He wandered a wartime Europe divided by religious questions, and over the course of his travels he became a freethinker. The astronomer tells us that Giordano wrote books, gave talks, was a teacher. His notions of faith, of an infinite universe, of the possibility of the existence of other lives, and the memory system he developed for the achievement of absolute knowledge, in addition to his questioning of the Holy Trinity and the authority of the Church, among other unsettling ideas at the time, spread, and stirred unease in the ecclesiastical establishment, which began to persecute him. In Geneva, the Calvinists arrested him and forced him to make a public recantation. Then he went on to England, Germany, and Venice, where—the astronomer tells us—one of his own students denounced him as a heretic and turned him over to the Holy See. The Inquisition locked him up for seven years, subjecting him to many interrogation and torture sessions. At the end of this long process, he was sentenced to be burned at the stake. The astronomer in shorts tells us that this time, despite the demands of the Holy See, Giordano refused to recant. Thus, at the age of fifty-two, sick and tired, he set out with guards toward the Campo de' Fiori, where he would be put to death. Along the way he endured the public shame of the sanbenito, a humiliating garment worn by the condemned, and a long walk under the gaze of the angry, howling mob. When at last he reached the Campo de' Fiori, Giordano was bound to a post atop a heap of firewood, where he spoke his last words. You pronounce this sentence against me with greater fear than I receive it, he said. Then the executioner fitted him with the requisite gag and lit the blaze in which Giordano would burn just like one of the stars he so liked to observe.

When he gets to this part, the astronomer stops. Maybe it's the cold, I think, and being underdressed for so long; maybe wearing those shorts has finally caught up with him. But that's not it. The young astronomer is overcome by emotion and he's having trouble talking. In a shaky voice he says that the stories of Giordano and each of the twenty-six victims we're remembering today are the same. Arrests, torture, interrogation, sentences dictated by the Holy See, or, in Chile, by military tribunals conducting trials unworthy of the name. Parodies enacted as part of the fantasy of a conflict that never existed. An assault requiring its participants to wear combat attire and carry curved knives and machine guns, justifying lunatic sentences like banishment, life imprisonment, or execution, handed down in the caricaturish ceremony of the tribunals, with absurd accusations like treason leveled for supporting the previous government. This treason thing, says the young astronomer, is so much like heresy. It all comes down to thinking differently. Dying for thinking differently.

For a moment the young astronomer seems lost. This time he really can't go on. He looks down, takes a step back, another step forward. He sets down the microphone, picks it up again. He can't muster a coherent gesture; the best he can do is weep. Weep for Giordano. Weep for Mario. Weep for our twenty-six stars. All of us—men and women—look at him in silence. We didn't expect this. His tears infect us with old familiar sadness and suddenly, without realizing it, we're crying with him. Releasing a sorrow from somewhere inside us, anchored in our bodies. A common sentiment that we acknowledge, an invisible tie that binds us, informing us that we're nothing but fish tied by the tail to other fish. The rope joining us is our good luck charm, protecting us. We bear it so as not to be lost, so as never to forget the shoal we belong to.

We sit here for a while longer, together in this space-time parenthesis. Enmeshed in this lament beside a monument in the middle of the desert. In the dark without a road map. Ceremony halted, script on pause, microphone turned off because we need to weep. Probably this is what we came for.

From among the women, the wives, the sisters, the sisters-in-law, the daughters, Violeta gets up and approaches the young astronomer. Exhausted Violeta is the one who goes to give him a hug. Expert at consolation, she whispers something in his ear and it's as if all of us—women and men— can hear what she says. He nods. He looks at her and nods as she wipes his tears with her wrinkled grandmother hands, her hands that have combed the desert. This young man could be her son or her grandson. He's only a little older than Mario was when he died. And there they stand, entwined in a long embrace that we watch without speaking.

Two fish tied by their tails.

Above us is the waxing moon, a few stars in sight. I think about other planets orbiting other blazing suns I can see from here, far away in the past. I think about those other lives that Giordano was talking about. In one of them maybe Mario and Violeta grow old together in a house full of dogs. In one of them maybe another woman embraces a different young man in shorts. In one of them maybe a man is burned at the stake in the middle of a public square and a group of people gather to weep for him in the desert. And maybe someone looks up at the sky. And someone sees the stars and thinks they're bonfires, cosmic blazes, celestial statues raised there in the night like a monument to stubborn memory.

▶

I have a hard time believing, says Violeta.

I learned not to believe.

Some said they were tossed into the sea.

Put in sacks.

Bound and taken away in trucks.

What I wonder is whether they were really tossed into the sea.

Whether they really ended up in the water.

Ham-nia

Ham k' Win Saik, translated from the Selk'nam tongue, means those who have gone. In funeral laments the phrase took the place of the name of the dead, which could not be spoken. From an old Selk'nam song about the departed: I stand in the footsteps of those who are gone. I believe that I come from those who are gone. Who are here no more. Who have left and gone away. And so I sing to Ham-nia, to the western skies.

The little people my mother talked about, the ones who sent messages in code with the flash of their mirrors—this is who they are: Dwellers in the western skies. Those who are here no more. Those who have left and gone away.

Aries

On October 5, 1988, my mother and my grandmother got up early to vote. It was the long-awaited day of the national plebiscite to decide whether Augusto Pinochet would remain in power or whether presidential elections would be called. The constitution drafted by the military regime—a document that governed and still governs the country today[2]—stipulated that a public referendum would be held with just two choices: Yes or No. If Yes won, and the candidate proposed by the regime was approved, Augusto Pinochet would effectively be elected and would continue to serve as president of the republic for eight more years. Elections of senators and representatives would later be called, and the governing junta would function as a legislative body until the forming of the National Congress. If the No option won, and the candidate proposed by the military regime was not confirmed,

---

2. In July 2021, in the wake of the social revolt that exploded in Chile on October 18, 2019, a democratically elected Constitutional Convention began to draft a new constitution. Its members include women and men in equal numbers as well as representatives of Chile's Indigenous people.

Augusto Pinochet's presidential term and the offices of the junta would be extended for a single year. Before the end of that period, presidential and parliamentary elections would have to be convened.

For whatever reason, with the dictatorship in full swing, my mother and my grandmother believed in the legitimacy of the plebiscite. They were convinced that conditions were right and the polls could bring an end to all the years of terror. They weren't the only ones. Much of the country believed it too. For the plebiscite, voter rolls were opened and political parties were legalized again. The Concert of Parties for No was born, with seventeen parties calling on people to vote. For the first time, television slots were assigned to both sides and international figures championed the No option. Jane Fonda and Christopher Reeve—Superman himself—showed up on television screens in the heart of every home, saying it was possible, we could do it, we had help. The eyes of the world were on us, safeguarding this unlikely escape hatch offered by the military regime itself. And despite an ominous blackout the night before, despite rumors of the cancellation of the plebiscite, despite attacks of nerves that kept them up for hours shuffling along the hallway with candles and cups of tea, at eight in the morning my mother and grandmother were ready to vote.

My grandmother got dressed in her formal best. She chose the blue linen dress she wore for birthdays and a light wool coat that hadn't yet become moth-eaten in her closet. The most revealing detail, the best demonstration of the morning's importance, was the absence of an apron tied around her waist. I could count on one hand the number of times I had seen her without it. She was eighty years old, the age my mother is about to turn, and she had trouble with her legs, so that it was hard for her to walk. This obstacle didn't keep her from making it to the National Stadium (table fourteen, to be precise), waiting in line, chatting with the other women, presenting her ID, signing everything that had to be signed, and then going into the booth to mark her secret ballot, as she did whenever there were elections. And so she did on that October day. Her vote was No.

I was seventeen. Much as I would have liked to, I wasn't able to participate. I watched them leave that morning and I remembered the last time my mother and grandmother had voted, back in 1980, when the dictatorship held an earlier plebiscite to ratify the new constitution that had just been drafted. After the military coup, the previous constitution had been suspended and there was a desire to start from scratch and create a new institutional framework. And so a commission was created to prepare a draft. After years of work, the document was subjected to the verdict of the citizenry on September 11, 1980, a day when my mother and grandmother—very well-dressed that time too—went out early to vote.

Their mood was not optimistic, as it would be eight years later. I remember them being irritable, complaining about everything. There were no voter rolls, no legal political parties, the opposition was suppressed, and of course there was no television or radio time for anyone who challenged the regime. Voting was regulated solely by stickers on ID cards and ink stamps on voters' thumbs, so you could remove the sticker, wash your thumb, and go vote as many times as you wanted.

My mother and my grandmother mocked the irregularity of the process. A show, they said, another piece of theater by the military regime to pacify the uneasy international gaze.

The constitution of 1980 was approved by a large majority. My mother and my grandmother laughed when the final tallies were announced on television. It was a joke, they said, such a crude deception that all you could do was laugh.

It was under this new constitution that the next national plebiscite was set for 1988. The vote was an important milestone in the dictatorship's design for the transition to democracy. Some of my young friends didn't plan to vote. They thought it was a joke to imagine a path to democracy under the rule of the dictatorship. Some of them laughed at the naïveté, saying the dictatorship would never uphold a result unfavorable to the military junta. And if it did, if it genuinely accepted the option of proceeding to elections, the next step would be the establishment of a system with the legitimizing veneer of democracy, though it carried on in the same dictatorial spirit, exacerbating existing differences. A post-dictatorship, they predicted. Others said it could be a way out and we had to take it. They said the old-school Democrats understood this better than we did, which was why they had accepted the choice offered by the military regime. We should trust them, not drag our feet. We had to let them act, not get in the way. At seventeen I didn't even have the choice. I simply wasn't old enough to vote. With the benefit of hindsight, this excuses me but doesn't let me rest easy.

Today, thirty years after the 1988 plebiscite, my son is invited by the student union and his school's history department to give a speech at an October 5 commemoration ceremony. My son is a devotee of Chilean history, and his admiration for his history teachers is boundless, so he happily accepts the challenge. He is seventeen, the age I was when I didn't vote.

▶

Aries is the first constellation in the zodiac and it is made up of eighty-six stars. The brightest of all is Hamal, an orange giant fifteen times bigger than the sun and orbited by a single planet. With a little effort and imagination, we can see that those eighty-six stars trace the shape of a ram in the sky. In Greek mythology, this ram isn't just any ram—it's nothing less than the ram of the shining Golden Fleece.

Phrixus and Helle were the children of King Athamas. In one version of the story, they are twelve years old; in another, fifteen; elsewhere, seventeen. Anyway, their young lives were in danger because Ino, the king's second wife, was jealous and Machiavellian, like so many women of Greek mythology— which in this day and age seems dubious and wrongheaded, to say the least. So much did she hate her stepchildren that she spent all her time scheming to get rid of them. Zeus, king of the gods, saw that the children's lives were in danger and decided to help them flee. And so he sent them the fabled ram, which was gifted with reason and speech and able to travel through the air as well as over the earth, its golden pelt visible from afar. The young brother and sister, frightened by their stepmother's conniving and conspiring, mounted the magic beast and set off in hopes of saving their lives. However, as they were flying over the sea, young Helle lost her grip on Phrixus's waist and fell before he could save her. Her body sank beneath the waves and she was lost forever. The boy, inconsolable at his sister's death, continued his voyage to the land of Colchis, on the Black Sea. After arriving safely, young Phrixus sacrificed the beast in tribute to Zeus. And Zeus raised the bright ram, instrument of the children's salvation, into the sky and made it a constellation.

I'm reading an astrology manual, looking for information about the sign of Aries. What I find describes eager, impulsive people, full of life and passion, capable of bringing change to the present and new energy to those around them. They know how to persuade others to break down barriers so life can advance and evolve. Their element is fire, light-giving and transforming. They are the crafters of the future, the pavers of the way, the drafters of new beginnings. As Aries is the first sign of the zodiac, its energy is linked to the zeal of early youth, I read in conclusion.

One April afternoon in 2001, as the sun—our mother star—was apparently transiting the constellation of Aries, maybe crossing part of the ram's left horn, a corner of his shining tail, or a golden centimeter of his gleaming coat, my son came into the world. He has no memory of the moment, of course. He has only my story, his father's story, the story of all the people waiting for him that day. My mother, my in-laws, some aunts, assorted friends, a couple of nephews. As his head emerged into the world and he gave his first cry, a bottle of champagne was uncorked in the waiting room. Just like in my dream, his arrival was celebrated with a grand toast and everyone wished him the best of lives.

▶

To talk about October 5 and the No vote, we have to talk about a lot of things. Too many things, maybe. There's no way to do it without referring to what happened during the dictatorship. To everything that happened under the previous government, the Unidad Popular. To everything that was gained in the Transition and everything that was lost. To everything our parents and grandparents achieved under democracy. To all the ways our parents and grandparents failed us under democracy.

We owe them so much and they owe us so much.

▶

Jaime Guzmán, a young lawyer in the early days of the regime, was an important collaborator and champion of Augusto Pinochet, acquiring a reputation as one of the regime's principal ideologues. He never occupied a position of authority, but he always operated as adviser and counselor to the military junta. Early on he was asked to take part in the commission to draft the new constitution of 1980, to be voted on by my grandmother and my mother, out early the morning of that election day in their best clothes and bristling with irritation. As part of this body, young Guzmán became a kind of spiritual father, crystallizing his conservative views of the world in the drafting of the law. His fear of freedom of the press and the right of assembly; his rejection of abortion, divorce, different sexual orientations; his minimizing of human rights; his defense of the traditional family as the bastion of Western society; his zealous protection of private property, free enterprise, and capitalism—all are part of the spirit of our legal system and therefore our way of life.

Whenever he showed his face on television, my mother and grandmother would swear at him. It didn't matter whether we were eating or whether we had guests: the four-letter words flew freely, often accompanied by a dish towel flung furiously at the screen by my grandmother. He was odd-looking and his way of speaking was excessively proper. But despite his obvious eccentricities, as a child I couldn't understand my mother and grandmother's hostile reaction.

Guzmán was the most important intellectual of the Chilean dictatorship. He devised its justification, the arguments in its defense, and its long-term political strategy. He urged the military junta to remain in power rather than limiting itself to a brief, surgical intervention. He excused violations of human rights with the explanation that they were exceptional, temporary, and necessary in an emergency situation like the one under which the country would live for seventeen long years. The project's success depended on the vitality and force of the regime, which is why the *dictadura* couldn't become a *dictablanda*, as he put it, punning on *dura* (hard) and *blanda* (soft). To defend this position he came up with an argument that shifted responsibility for the violation of human rights onto the previous government. Guzmán spoke of a hypothetical civil war that Salvador Allende's government might have caused, and he declared that there had been a collapse of democracy prior to the military coup in 1973. According to this logic, the military had simply responded to the Allende government's push to put the country on the path to totalitarian Marxism, in his words. In this fantasy of belli-cose engagement and deluded portrayal of nonexistent com-bat, the war councils, banishment, execution, torture, forced exile, indefinite detention, interrogation, and disappearance of people looked justifiable. Buoyed by this false narrative of civil war, Guzmán and the political party he founded claimed democracy as the rationale for their participation in and un-conditional support for the civil-military regime.

As a little girl, I thought my grandmother and mother knew him personally. I imagined he was some distant rela-tive, a former neighbor who had done them an unspeakable wrong, something terrible and dark to justify the hatred hurled at the TV screen. In time I realized I wasn't entirely wrong, and in fact Guzmán, whom my grandmother and mother had never met in person, had meddled not only in their lives but in the lives of everybody else in the country, including mine. Including my son's. So radical was this meddling that when I finally understood it, I joined in the family ritual. There I was, yelling and throwing dish towels at the television until the day Guzmán was shot and killed at the entrance to the university where I was a student.

The best tool he crafted for the perpetuation of the regime was the constitution of 1980. It's a constitution safeguarded by rules intended to limit any change, block any exit, prevent any escape. A constitution that—as I've said—stipulated the transition to democracy just as we experienced it: with a plebi-scite held on October 5, and subsequent presidential and par-liamentary elections granting the dictator Augusto Pinochet a seat for life in the senate when he was ready to give up his role as commander in chief of the army.

With this history in mind, and as inheritor of the family ritual inspired by Guzmán's televised face, my son wrote his speech for the October 5 anniversary and shared it with his classmates on the student council.

▶

During the dictatorship Chileans couldn't express their ideas if they were different from the regime's. During the dictatorship people couldn't publicly express their sexuality if it was different from the regime's approved version. During the dictatorship people were afraid to listen to a protest song or buy an opposition magazine. Thousands of Chileans were killed, tortured, and made to suffer just because they thought differently.

▶

The text was then critiqued by the student council. There was a discussion about points that could be refined, and a consensus was reached on structural changes to enhance the speech before it was delivered the next day. My son—let's call him D from now on—was satisfied with his classmates' feedback and returned to the school routine. Note-taking, tests, breaks, snacks. More note-taking, more tests, more breaks, and more snacks. In the middle of the whirlwind of school, class was interrupted by one of the teachers from the history department—we'll call her A. She came in and asked permission to speak privately with D, leading him into a corner of the courtyard hallway.

Sitting on a bench in the middle of this no-man's-land,

A thanked D for writing the speech for the October 5 ceremony. She said she was sorry she hadn't had time to read it in full. That's teaching for you, she said. Constant work and never a spare moment. And after she apologized, out of nowhere she asked D to cut or rethink three particular sentences before giving the speech the next day.

D listened in surprise to his teacher's request. As I've said, he had just had a conversation with his classmates and together they had settled on revisions so that the speech duly represented the voice of the student council at the ceremony. He hadn't realized he was also expected to coordinate with the teachers. Anyway, if A hadn't had time to read the text in full, how could she ask him to cut certain sentences?

A persisted. The ceremony had been organized by the history department and any remarks would have to represent the teachers too. There was no way students and teachers could have different views on the subject. This struck D as strange, because wasn't it natural and healthy—almost a given—that students and teachers would have different views on everything, or almost everything? Also, the students hadn't been shown any speech to be given by the teachers. If the adults were going to weigh in on what the young people had written, it seemed only fair that he and his classmates should weigh in on what the teachers had written.

This meeting in the hallway was now joined by another teacher from the history department. We'll call him B. B wanted to add his voice to the conversation. He made it clear that he, too, was familiar with the speech and that he very much agreed with A. As things stood, the three sentences in question made the text hurtful, hostile; it fell into the same prejudiced and intolerant habits of the dictatorship it discussed. The idea—the history teachers said—was that no one should feel left out at a ceremony commemorating the arrival of democracy. His speech should therefore be written in that inclusive spirit.

D tried to justify the three hurtful, hostile, and intolerant sentences that he was being asked to cut or rephrase. But every defense he offered was countered by his history teacher with arguments about practicing tolerance, and so he began to consider the idea of revising the text and bringing in a new draft the next day. In encouraging tones, A and B emphasized that any questions or comments on his speech should be taken simply as suggestions.

▶

My mother reads what I've written about Jaime Guzmán on the computer I've left on. She asks me whether she really yelled at the TV whenever he made an appearance. She remembers how upset he made her, the anger she still feels, but she doesn't remember arguing with the TV or swiping at it with the dish towel. I try to re-create a scene with stray bits of memory, and I feed her details, hoping to land on something familiar.

The living room table with its flowered vinyl tablecloth, cups of tea, maybe plates and silverware for elevenses or dinner. My grandmother's chair pulled close to the television because of her nearsightedness, and the old black-and-white TV set with an antenna that had to be wiggled whenever the picture went funny. One night like any other, probably during the eight o'clock news, Guzmán comes on-screen for an interview. A brief introduction of the topic of the day, maybe a question from the reporter, and then that barrage of insults drowning out whatever they're talking about. And as a finale, the glorious hurling of the dish towel, which, with any luck, hits the television set.

My mother laughs at the story. Her face lights up; she says it may very well have been that way, but she can't remember. She really can't remember.

For a moment I wonder whether the story I've told is real or something my fanciful mind came up with on its own. Seeing how surprised she is, I begin to think I'm the one imagining things. I do a quick scan of my brain, a general review, and as I sift through my mental snapshots I confirm there is something untranslatable, some bodily certainty, some instinctive radar that indicates when something is a real memory and when it might be a trick. I could be mistaken, but I would swear by this memory. It was a ritual repeated over and over for years, a bit of my history, a little piece of me. To disown it would be to disown my own hand, my ear, my navel.

In April 1991, as the sun—our mother star—was apparently crossing some part of the constellation of Aries, Jaime Guzmán was assassinated. Astrologically speaking, you'd have to say his death happened under the ruling influence of the ram, a passionate, fiery force unafraid of breaking down barriers so life can advance and evolve. It happened late in the afternoon. He was on his way out of the university where he taught. He was a senator of the republic, still a key political player in the newly formed democracy or the beginning of the post-dictatorship that my young friends—so very young—had foreseen. The constitution he crafted had made it possible for Guzmán himself and many Pinochetistas to be part of the National Congress. On this count, his car was fired upon at the campus exit that afternoon in a well-executed operation. His driver took him to the headquarters of the political party he had founded and then to a hospital, where he was given emergency care.

A few minutes after the shooting I exited the same campus with a friend. Tensions were high at the entrance to the university. Something had happened: shots had been fired; there was a fleeing car, people screaming and running, broken glass on the ground, a few national police officers arriving on the scene. Nothing was clear and nobody could explain what was going on. Probably nobody quite understood what had happened or believed it yet. Assassination attempts were unsettlingly familiar and our sensitivity to shootings was low. After all, we had grown up amid gun battles. I guess that's why neither my friend nor I was upset by the confusion. I'm trying to remember more details, but I can't. We were late for a theater rehearsal at my house. We didn't have much time to lose, so we decided to keep walking and get on the bus. Maybe we thought we would find out more the next day; someone would tell us what had happened. Half an hour later we found my mother and my grandmother in front of the same old black-and-white television set, watching the news. Jaime Guzmán was at the Military Hospital, being operated on in an attempt to save his life. His family, his fellow party members, Augusto Pinochet himself were gathered there. They all filed across the screen as my mother and my grandmother watched in silence. There were no insults this time. No dish cloth was brandished. Possibly a fleeting, imperceptible smile settled on our faces. A timid but brutally honest smirk. Or maybe not. Maybe I'm making it up. Anyway, we guessed what was coming. A few hours later Guzmán's death was announced on the screen of that same old television set. That evening our ritual came to an end.

Some of the images I've summoned have hidden themselves away in a parenthesis of my mother's brain. The same place everything ends up when she faints, maybe. All those seconds she loses before fading to black, reconstructed by a chorus of generous voices. You hugged the wall, you held your head, you vomited, you sat down on the ground, you closed your eyes, you collapsed. There is so much material in there, years and years filed away. I suppose it's hard to organize it all. These memories of Guzmán—she can't find them anymore. Now they live only in my memory and in this story as I tell it and write it.

I remember the electrical charges I saw in her neurological exam. Those constellations of clustered memories. And I muse, in a rather obvious way, that the parentheses in her brain are like the black holes of the cosmos. Dark, enigmatic spaces packed with hidden information. I have only the most basic understanding of them. A massive star uses up its nuclear fuel, and instead of collapsing outward, shrinks to an unimaginably small point of infinite density. This point has a border called an event horizon. A kind of boundary between itself and the outside. Here the force of gravity is so strong that light is captured and cannot escape. Since nothing travels as fast as light, it goes without saying that anything else would be dragged inward too. Nothing can possibly emerge. Whatever falls in is lost forever.

If I think about the lived moments my mother can't remember anymore, if I think about the moments I've forgotten myself, I'm reminded of those black holes. I imagine that my lost memories are trapped in some mysterious part of my hypothalamus, trying to take up as little space as possible, so carefully tucked away that they become imperceptible. Ghosts. Presences without substance, without color, without memory. Secret stars that are there, though we don't know it.

If I think about the story of Mario Argüelles and his twenty-five fellow victims executed in the desert, if I think about all the people of Calama, their city, who have no information about them, I'm visited again by the image of those menacing black holes. Twenty-six lives and twenty-six deaths and twenty-six bodies hidden in some corner of history, in a blind spot where there's nothing left to be found anymore.

I think about how we tell the stories that make up our lives. I think about how we choose what to put in them and which point of view to tell them from. I think about myths. Those foundational tales that orient entire civilizations. I think about the way they frame archetypes, thoughts, and actions. I think, for example, about poor Helle, who lost her grip on her brother Phrixus's waist and fell into the sea. Some arbitrary decision erased her and not him from history. What would have happened if it had been Phrixus who fell into the sea? What would have happened if Helle had survived? I also think about those wicked, jealous women like Hera, Gaia, and Ino, capable of plotting the most brutal crimes against their husbands' innocent children. I think about this dubious and wrongheaded constant in the plots of myths and about the limited, unfair female image rooted in them. I think about the great narrative of history. About how it's told. About all the biased and manipulated information. About paradigms raised as flags. About fabricated wars. About the fictions devised to govern societies, countries, eras. I think about how our lives are controlled by these arbitrary and even absurd fictions. Whole generations unknowingly playing parts in a script written by a select few. We repeat the same plot over and over again. They teach it to us and we teach it to others, teach them to follow the path over the cliff without realizing it. We might even go so far as to uphold and defend that script. We're capable of spending whole lives following the rules made for us, resisting change, going in circles, treading on our own tails like rats in a lab, never imagining there are other possible realities. I think about our own event horizon, how it is drawn for us. About everything that crossed over and fell into the void, everything that was absorbed by a dark force and lost, everything left without a place, now and forever. The excluded names, the invisibilized groups, the hidden horrors, the redacted opinions—and then again comes the vision of those terrifying, menacing black holes.

I used to believe they were empty space, blots of nothingness lying in wait. Now I realize they're actually places of great density of information, of material maximally condensed until it's no longer detectable. As if some cosmic law were censoring the content of stars, black holes hide a message rendered invisible to our eyes.

But just because we can't see it doesn't mean it doesn't exist.

▶

Today, thanks to October 5 and the No vote, young people can express their opinions freely. We can declare our diverse sexual preferences, our passions, our political opinions, and not be afraid of what might happen to us. Thanks to October 5 we can vote, choose who governs us, be part of the democratic process. We can march and express our differences.

But talking about the No vote doesn't mean only reflecting on how far we've come. Talking about it means remembering, and recognizing how far we still have to go.

▶

On October 5, 2018, thirty years after the plebiscite, D went to school, excited to give his speech. During his first-period class there was a conversation about the issues, with the teacher leading the discussion and all the kids sharing their feelings. There was talk about the importance of commemorating the day, the importance of democracy, the importance of voting, the importance of remembering. And in the middle of this exercise of great importance, the president of the student council came into the room and asked permission to speak to D. Excused from class, D followed his classmate into a corner of the hallway. Here, an hour before the ceremony and the presentation of the speech, the student council president told D that the history teachers had come to him to express their concerns about what he had written. They still believed the text couldn't be read as it stood, with those three hurtful, hostile, and intolerant sentences. And so the young leader wanted to know whether D had taken them out or reworked them in the new version.

D explained that he had made the structural changes agreed upon with the student council, and he had thought a lot about his history teachers' suggestions. But after trying to reframe those three sentences, he had come to the conclusion that he couldn't do it, because any attempt to soften what he had written or play it down would lessen the impact of his ideas. And it was important to D to be faithful to the substance of his ideas.

The student council president—a young man the same age as D, a dear friend of D's—said that after speaking to the history teachers he had noticed the same thing about those three sentences. Though they hadn't seemed hurtful, hostile, or intolerant when they'd talked the day before, now, in the light of his teachers' observations, he'd realized they were. There was a radical tone to them that bothered him. Such harsh opinions might upset more than a few people.

Before the conversation could go any further, B stepped into the hallway and asked the two boys to come with him. D and the student council president walked to an office where a third teacher from the history department was waiting for them—a person to be called C. B and C were prepared for the conversation and they had a printed copy of D's speech on the desk with the three hurtful, hostile, and intolerant sentences underlined in red. D could see his words swimming in colored ink. It looked like a marked-up assignment, unsatisfactory work that would clearly get a very poor grade.

Minutes before the ceremony, with little time to lose, the

history teachers again stressed how uncompromising and biased his speech was. The stance he had taken—declaring who should be allowed to express themselves and who shouldn't—had a lot in common with the position of the dictator he condemned, and that was no way to proceed in a ceremony celebrating the arrival of democracy. Those three sentences had to be reconsidered.

There was no point mentioning Jaime Guzmán in the speech. D might not believe there should be streets named after him or monuments erected to him under democracy, but it was irrelevant, because the kids at the school didn't even know who he was. Why bring up a historical figure no one knew anything about? the history teachers asked. They didn't see Guzmán as some sacred figure—he was a questionable character, no doubt, but personal opinions had no place at this celebration.

It was the same with the second sentence, the history teachers said. The one about Patricio Aylwin, first president of the Transition. To say that he applauded the military coup, even if it was an established fact, was unnecessary, because it was hurtful, the history teachers said. Any evidence of complicity between the first president of the Transition and the military coup should be omitted. Avoided. A speech celebrating the arrival of democracy should be an inclusive speech, representing all views, seeking consensus, not leaving anyone out.

Also unnecessary was the point made in the third sentence in need of revision, said the history teachers. About how it was unacceptable for political parties that had been active in the dictatorship and still defended it to continue to exist. Of course it was a regrettable fact—they felt the same way—but one couldn't outlaw the views of people who belonged to right-wing parties. In democracy we all have the right to our opinion, they said. Freedom of expression should be defended and differences should not be suppressed.

Forty long minutes went by while the history teachers and the student council president tried to make D see reason and come around to their sensible, conciliatory, inclusive point of view. They tried in many ways to convince him to cut the three sentences and forget about them.

D once again did his best to defend his position. They were misunderstanding him, he explained. To him, a right-wing political party wasn't the same as a party that had been active in the dictatorship and that continued to defend it. There was a fundamental difference between the two. His question was why, thirty years after the plebiscite, the Pinochetista right was still playing a leading role in the country's politics via its parliamentary representatives, mayors, and opinion leaders, the great majority of whom had been part of the government under the dictatorship. As a young person, he couldn't see how it made sense to celebrate October 5, to talk about its importance, if we weren't going to harness the meaning and energy of that memory. Why commemorate the arrival of democracy if society continued to allow profoundly undemocratic opinions and movements to flourish? He didn't understand why he was being asked to be respectful when referring to people who had supported and worked with a regime that didn't respect differences, that tortured its opponents, that made people disappear, that exiled or killed those who thought differently. That's what he said to his history teachers. He didn't understand how democracy could fight for itself unless those who believed in it stood up and took action. One way to take action was to set boundaries to protect it from those who opposed democracy. Intolerant rhetoric could not be tolerated. There could be no paying homage to those who were complicit in the brutality, as Jaime Guzmán had been. It was hard to imagine the kind of uproar there would be in Germany if someone decided to erect a monument to Joseph Goebbels, for example, or even suggested naming a street after him. The democratic expression of ideas and differences was one thing, but permissiveness toward those who had installed a regime by systematically executing its detractors as a matter of government strategy was something else. Not declaring this openly seemed irresponsible and dangerous to D. And it was he and his generation—young people who were being kept in ignorance, forgetful and unquestioning of the past—who would pay for this deed of omission, he said. They would be the scapegoats left to usher in the déjà vu of History that was already making its presence felt in the new fascisms appearing around the world.

D was under pressure to retract his words even before they were spoken. He understood that the cuts proposed were in no way only suggestions. The inclusivity and tolerance of the democratic celebration of October 5 didn't apply to him. There was no reasoning with his history teachers and the classmate who had changed his mind and now agreed with the teachers—the best proof, so they said, that they were right.

D felt the weight of the ages on his slight seventeen-year-old shoulders. The power of a dangerous ancient script he didn't want to follow. The attraction of a black hole trying to suck in his ideas, make them invisible, trap them, and hide them in a dark corner. It was as if a mythical and unjust punishment had befallen him. Born under the sign of the ram—eager, impulsive, full of energy and passion, crafter of the future, capable of bringing change to the present and new energy to those around him—D realized he was alone. No Golden Fleece would come to his rescue when history repeated itself because of the tolerant and conciliatory attitude he was being forced to adopt. There were no responsible adults, or so it seemed. Probably there never had been.

▶

On this fifth of October, I would love to talk about how far we've come, how well our democracy works, but it isn't easy.

How is it possible that political parties that were active in the dictatorship and continue to support it in part or in full still exist? How is it possible that political leaders in parties that worked with Pinochet are members of parliament today?

How is it possible that there are public places named after leading figures in the military regime, like Jaime Guzmán? The existence of streets or monuments bearing his name is a celebration of the greatest civilian accomplice of an authoritarian and genocidal dictatorship.

How is it possible that there are mothers and fathers who still don't know where the bodies of their sons and daughters are buried? How is it possible that they continue to die without knowing? How is it possible that some of the people responsible for these crimes are free to walk the streets or even have generous pensions paid by the state?

~~How is it possible that we're surprised by the fact that the first Transition president was involved in the coup? How is it possible that we fail to see the democratic ethic this promotes?~~

How is it possible that education and health are consumer goods, not rights? How is it possible that our grandparents lack respectable pensions? How is it possible that we live under the same constitution drafted by the military regime? We'll never have a true democracy so long as we hold on to that illegitimate constitution. It doesn't matter how much we patch it up. It was illegitimate from the beginning.

I would love it if we could truly express our sexuality freely and there were no homosexual or trans people being beaten in the street. I would love to see my female friends go out at night and not have to fear any man. I would love to be able to march without witnessing how the police beat us, poison us with tear gas, and arrest us unjustly.

But nevertheless I should be grateful that I'm able to be here, reading these words to you. I should be grateful that I'll suffer no consequences for what I'm saying. For expressing everything I truly think.

This is why we must remember and thank those who had the spirit, courage, and solidarity to vote on a day like today thirty years ago. With the power of the pencil, they achieved what seventeen years of armed struggle could not. But we must also thank those who fought the dictatorship with other kinds of weapons. All of them—women and men—deserve our respect.

We don't celebrate a day like today in order to congratulate ourselves, but rather to give ourselves the push we need as a coun-

try and as young people to advance toward a more just, free, and creative future, clear-eyed about the past.

These words are for everyone who fought during those seventeen years and for all those who are still fighting. May the strength of their example lead to happiness once and for all.

D.L.F.
Santiago de Chile, October 5, 2018

▶

As if some cosmic law were censoring the content of the stars, black holes hide a message rendered invisible to our eyes.

But just because we can't see it doesn't mean it doesn't exist.

www.constelaciondeloscaidos.cl

Gemini

Please, let me be.
Let me continue my voyage to the stars.
—Golden Record, Voyager 1 and Voyager 2

There are eight candles burning on my mother's birthday cake. Small torches flickering delicately atop the chocolate-and-cream icing. One for each decade lived. How much life, I ask myself, is contained in each candle? How many images, faces, smells, tastes? How many moments? A constellation of gray-haired women accompanies my mother in this ritual. These are the friends who helped me organize this surprise party. They are part of the fire; their lives are bound up in the sharing of this blazing cake. Secret memories, mutual understanding built up over many years, hundreds of shared moments maybe already consigned to some mental black hole, though they still have a place in this cake, because what we forget takes up space just as surely as what we remember.

My son's father insists on a photograph. We all smile at the camera. We hold on to this instant with an image that in the future will help light the flame of some other cake. A chorus of voices sings "Happy Birthday" as the circle of candles lights up my mother's tired green eyes. In her singular head she harbors eighty years of conscious recollections, hundreds of thousands of neuronal constellations combining and dredging up memory, reverberating in there, putting on the most incredible light show. And despite the energy expended, despite all this activity in her brain, my mother closes her eyes before blowing out the candles and seeks a place to store one more memory.

She secretly makes three wishes: Not to forget this moment. Not to forget this moment. Not to forget this moment.

▶

Beyond photograph and memory, where will we end up? What skill will be needed to pierce the layers of time burying this instant? Where will the laughter of these women go? The smoky smell of blown-out candles, the chocolate crumbs on the white tablecloth? Will they be recycled somehow? Will they turn into dreams? Will they drift lightly like a kite, falling when we least expect it? What will happen to my mother's tired green eyes? What will happen to her twisted hands? To the fine gray hairs she leaves on the back cushion of the armchair? Will memory be able to recover it all? Will an exact copy be kept, to be resorted to when needed? A clear script so that nothing is forgotten: the voices, the hairdos, the smells of each body, the lulls in conversation? Will this moment be replayable, at least one more time, in someone else's brain?

▶

Like all bodily organs, the brain has evolved, growing increasingly complex over the millions of years of human existence.

The brain with which our ancestors etched the phases of the moon on a bone, devised the zodiac in Babylon, or tracked the stars in the middle of the dark desert can't be exactly the same as the brain of this woman blowing out her birthday candles, overcome with emotion.

In a book on neuroscience I read that the current structure of the brain reflects all the phases it has gone through. The history of its development is there, embedded in its architecture. At the base is the brain stem, which manages basic biological functions like heart rate and breathing. This is our reptilian brain—the seat of instinct, aggression, ritual, territorialism, social hierarchy—as it evolved in our reptile ancestors. In the depths of our skull, we house something like the brain of a crocodile.

Surrounding this is the paleo-mammalian brain, which evolved tens of millions of years ago in our mammalian ancestors, before primates existed. This is the site of the limbic system, the neural network that weaves our emotions and syncs the basic survival processes of the reptile brain with the world. Fear, rage, happiness, and sadness circulate here—the full emotional gamut that rules our moods.

The layer over these more primitive brains is the neo-mammalian brain, which evolved millions of years ago in our primate ancestors. It's here, in the cerebral cortex, that our consciousness resides. This is our space for sensing, reasoning, drawing conclusions, and giving meaning to things. This is the brain that lets me consult books on neuroscience, process information, and file it away in my memory. In this realm, we learn from what we read, what we experience. We form abstract thoughts, pretending that the cosmos and the brain are two similar mysteries and that stars and neurons have a secret connection. This is the brain that lets me understand how the brain works, more or less, how it developed and expanded over time as required, like a city, retaining its initial structures and building others around them to fit new contexts. The reptile brain is like the Plaza de Armas in Santiago de Chile. Ground zero, foundation of the urban grid. Site of the first executions, of commercial exchange. The brain, like the city, has spread outward from its old center of operations. But unlike Santiago, it has followed an urban plan that values its heritage. It has honored its past as it grows, protecting it, using it as the fundamental base for development.

We are containers of memories. Our cerebral structure and our genetic makeup prove as much. But while the DNA in our genes is a site of permanent memory, remembering is only part of what the brain does. Our neurons create and store content beyond what is genetically inherited. From the moment we utter our first cry at birth, our desire to learn is the fundamental tool for our survival and development. Thanks to the work of the cerebral cortex, we harvest the genetic tales that are inherited from our ancestors and carried in the body, and we create new ones in an unending operation of memory and action, past and present, yesterday and today, bringing about the development of whole civilizations.

Where our genetic memory couldn't store any more information, the brain stepped in. When we realized we were dealing with more information than the brain could handle, we invented ways of keeping records, and thus writing arose, for example. The first writings were chiseled on clay, stone, or bone, invented to keep track of practical information. Symbols and drawings aided in measuring grain, fixing the position of the stars, identifying the phases of the moon and the change of seasons. Later, people wrote on bark or on leather; they painted on papyrus, bamboo, silk. Enigmatic hieroglyphics gave way to particular alphabets, symbols that sought to translate the sounds of each language system. It was a long time before rice paper and ink were invented in China. A long time before the arrival of a printing method using intricate pieces of porcelain to stamp characters, which made it easier to produce copies of texts. Some years later, printing as we now know it arose in Europe. This made it possible for texts to be distributed in book form, passing from hand to hand, place to place, and era to era, so that information spread and advanced into the future.

A book is a space-time capsule. It freezes the present and launches it into tomorrow as a message. Our ancestors recorded their surroundings, reflected on them, made surprising discoveries. They gave voice to their preoccupations, their beliefs, their gods, their wars. They told their own stories, invented others, erected fantasies, built parallel realities, legends, myths, songs, poems, fictions, novels—and of this and more they still speak to us from the pages of books. We can make contact with the most sophisticated minds, the thoughts and sensibilities of history's great teachers. Like shepherds following the stars in unpolluted places or the ancients charting their way across the desert at night, we let ourselves be guided by the messages left behind in these writings. We are inspired by them; they play a part in shaping our present and our future. On the astral map of books we are bound by invisible threads in a relay of wisdom and knowledge, content and images, light and dark, that began long ago and on which we are just a way station.

A piece of sky can be a library too. Memories travel, breaking the laws of time to reach our hands, our eyes, our minds. Our lives.

As far as we know, we're the only species on the planet with the need to accumulate memory outside our brains. Drawings, paintings, recordings, photographs, films, books: Each of these platforms that eventually became art originated as an attempt to hold on to reality and pay tribute to our surroundings. Instants captured and eternally preserved, becoming part of the past the second they are fixed. Sounds, images, voices, exhalations, faces, thoughts, reflections, ghostly landscapes—forever reliving a piece of yesterday. Recorded and saved to be found again. The hunt is always on to rescue them from oblivion and add them, like stray puzzle pieces, to the broken mirror in which we've always tried to see ourselves.

▶

The *Voyagers*, twin exploratory probes, were launched by NASA in 1977. Their mission was to observe the outer planets of the solar system, due to align at the end of the 1970s. Jupiter, Saturn, Uranus, and Neptune could all be visited and observed in the span of a few years, inspected by two travelers setting out days apart on similar courses that in time would completely diverge.

Looking at a photograph of the twins, we see that they resemble a couple of insects. Assorted arms and antennae jut from their bodies, making them look curiously like cosmic bugs. Their sophisticated structures house cameras, light sensors, and radio astronomy equipment, as well as instruments for measuring and interpreting temperature, color, plasma waves. They bear multiple devices for the detection of particle energy or the composition of celestial bodies. The *Voyager*s are equipped to be two perfect huntresses. Their work is to record and store cosmic moments. To function as a pair of artificial brains gathering fragments of stellar memory. On their voyage they discovered new moons, unheard-of volcanic eruptions, an ice crust on Europa, craters of all sizes on Callisto, a ring of delicate particles around Jupiter, high-speed winds on Saturn, surprising cloud formations on Neptune. And they kept a record of each of these moments. Postcards bringing news to Earth of their extraordinary voyage. Thousands of snapshots of landscapes to which no woman or man has ever had access.

Voyager 1's encounter with Saturn flung it far off its original course. The same happened to *Voyager 2*, sent in the opposite direction by Neptune's gravity. Thus the twins were separated once and for all, abandoning their recording work and beginning a new mission. Now they would be leaving the heliosphere, crossing the boundary between the solar system and interstellar space.

The twins are beginning to run out of power on their voyage. To conserve what they have left, most of their instruments have been turned off. In February 1990, *Voyager 1*'s cameras were used one last time to take a photograph of the solar system from afar. Shot from six billion kilometers away, the images are the conclusion to a yearslong process of record-keeping, the *Voyager*s' farewell to Earth, one of a gaggle of planets, a tiny blue dot in the vastness of space, picked out from a great distance by a ray of sunlight.

Carl Sagan, astronomer and science communicator, creator and host of the popular television series *Cosmos*, freethinker, skeptic, atheist, writer of endless papers and books, was partly responsible for those last photographs. Sagan, who participated in the *Voyager* program, suggested pointing one of the twins' cameras at Earth to capture a final distant image. From the resulting series of photographs, the single shot in which our planet is captured is still the most faraway picture ever taken of it. Inspired by the photograph, Sagan declared it proof that we were just "a mote of dust suspended in a sunbeam."

At the beginning of the 1980s, Sagan's face appeared on our old black-and-white television, his pleasant-sounding dubbed voice supplied by that of some Mexican actor, to introduce and present a new episode in his series. Once a week for an hour, in those strange times in which I was fated to grow up, I sat down eagerly in front of the screen to escape the tight bounds of my small world and travel through time and space to the accompaniment of the series's Vangelis score. Each episode was an adventure that led me to different eras, places, and knowledge. Astronomy, history, science, biology, the origin of life, human development, space travel, the stars, the infinite universe, the awareness of inhabiting a planet that is nothing more than a mote of dust suspended in a sunbeam.

As helicopters flew over the roofs of my neighborhood, as we lit candles to see by during each blackout, as we tried to survive the pull of the black hole that sought to plunge us into the deepest ignorance and deceit, there was the certainty that every so often, for fifty minutes or more, we could set out via the screen on a voyage of discovery. An escape hatch to another possible reality, far from gun battles and curfews. On those television adventures I learned that the current moment was insignificant in the cosmic scheme of things, that there were infinite points of view on any subject, and that all knowledge is the fruit of deep, relentless questioning. And though Sagan was talking about science, I felt his words were a secret message for me, a South American kid trying to understand the broken country where she was fated to live. He invited us to doubt everything. To mistrust truths and constantly interrogate our surroundings. To reject official narratives as well as blatant ignorance, nonsense, and lies. To look beyond our small piece of ground with all the powers vested in our brain, which for millions of years had worked at evolving to make such a thing possible. To leave our own body and fly over house, neighborhood, country, world. To slip past borders and observe from on high, our gaze panoramic, encompassing every point of view. To be an exploratory probe, roaming the universe and deciphering its laws.

All these years later I'm in front of a screen again, watching an episode in the series and realizing that the words I heard as a girl are still relevant. The memory of them ignites in my mind and illuminates our unhappy present. This perfect blend of power and ignorance has existed since the beginning of time in a historic continuum that is as yet unending—ask Giordano Bruno; ask Mario Argüelles. Humans have evolved to the point of having brains capable of expanding into artificial brains, of traveling into space and writing exquisite, inspiring poems, and yet stupidity still reigns. Every day we're bombarded with the most absurd declarations meant to keep us on the event horizon of the vast black hole we're about to topple into. We're told that global warming doesn't exist. That global warming is good. That global warming will lead to new business opportunities and possibilities for mineral exploitation. That cancer can be cured with the toxins of an Amazonian tree frog. That Brazilian monks do their work from afar with offerings of glasses of water and candles lit for them. That walls must be built to divide nations, that they must be wired for electricity and rigged with cameras. That some races are better than others. That some countries are better than others. That Earth is flat and that any other explanation is just a centuries-old international conspiracy. That the Holocaust never happened and that any proof is part of another soon-to-be-centuries-old international conspiracy. That it's a good idea to cut back on history in school curricula. That it's even better to cut back on philosophy. That homosexuality is an invention of degenerates. That transsexuality is an invention of other degenerates. That boys should wear blue and girls should wear pink. That women are meant to be mothers and mothers only and that they should not be allowed to make decisions about their bodies. That all women are jealous and vengeful and that they plot to torment men. We're told about terrorism, delinquency, security, keeping the peace. About Father Christmas, the tooth fairy, the bunny with the chocolate eggs. About God, the Virgin, and the Holy Spirit, who forgives and saves us all.

We need to sit down in front of a screen again and make more frequent trips to the stars. We need to heed their warning messages. We need to look again and again at that old photograph of Earth taken by the twins from millions of kilometers away and consider the fact that the mix of power, ignorance, stupidity, and nonsense that has spread disaster for centuries on our planet has been motivated solely and exclusively by the desire to dominate a small pixel of an insignificant dot in the cosmos.

Nevertheless, despite that whole campaign to blind us, the *Voyagers* are moving through space at this very moment, traveling with their golden records bearing our attempts to register the best of humanity. A kind of greatest hits rejecting all ignorance and stupidity. A distillation of our best selves, our noblest essence. Phonographic records complete with cartridge, stylus, and instructions for use, intended as a means of contact if they're ever found by representatives of another civilization of the cosmos. Something like a letter of introduction.

Nine months before the twins' launch, NASA asked Sagan himself to take charge of this message. The idea was to put together a sampling of the diversity of Earth's cultures and lifeways. The chance that another being in the universe would bump into the twins in some distant future seemed slim (and still does), yet the small team took the challenge seriously and began to work on the selection of materials. But how to curate the experience of life on Earth? What guidelines to follow? How to justify the time and energy spent?

After much thought and consideration, the team decided that the records would contain a selection of the planet's best music. Pieces from a host of cultures featuring the voices and instruments of the most diverse performers. Bach, Stravinsky, Beethoven, Mozart, Chuck Berry's "Johnny B.

Goode," Senegalese drummers, alongside Bulgarian chants, Peruvian panpipes and drums, and much else. They also chose to include greetings. Words intended to send a message of friendship. Though the potential recipients might not understand any of it, they recorded messages in nearly sixty languages, from Sumerian (the oldest tongue on record), to the calls of humpback whales, to the words of a five-year-old boy from the United States. And a panoply of voices in Aramaic, Syrian, Mandarin, Nepalese, Spanish. Sentences spanning the gamut of recordable languages.

Hello, good wishes to you all. / How are you, people on other planets? / Friends in the stars, may we meet someday. / Long life to you and everyone in the universe. / If there is someone out there, we wish you all the best. / Please, be in touch. We're here; we say hello. We are the people of Earth.

There are photographs of women and men, too, and scenes from across the planet. Houses in Mexico and Africa, a street in Pakistan, a jam-packed road in India. Views of the Taj Mahal, the Great Wall of China, the Sydney Opera House. A woman looking into a microscope, a Balinese dancer. Thai craftspeople, mountain climbers, gymnasts, an astronaut in space. An elephant, an airplane, a train, the X-ray of a hand, the inside of a museum, a human heart. Geological sounds were included in an attempt to give a sense of the origins of life on Earth. Rain, wind, the crackle of fire. The shriek of a chimpanzee, a chorus of birds. A flock of sheep grazing, the bark of a dog. Also included were sounds illustrating the vertiginous development of new technologies. Trains, cars, a rocket taking off. Added to all this material were the thoughts and feelings of a human being: the electrical activity of a human brain, recorded and transcribed into sound, then compressed and added to the record.

This human being registered in the twins' memory was a woman. Her name was Ann Druyan, and she was a member of the team that chose the contents for the *Voyager* records. It was her idea to include an electroencephalogram among the materials. Since EEG patterns show changes in thought, maybe they could be deciphered if found. The team agreed, and on June 3, 1977, Ann underwent the EEG procedure herself, allowing the electrical activity of her brain to be recorded.

I think about this, and neurons constellate in my brain, bringing back my mother's neurological exam. That hospital room, her agitation, her bewildered eyes. Like my mother a few months ago, I imagine Ann must have lain down on a hospital bed, her scalp covered in electrodes. Surely she, too, must have followed the instructions given by some doctor. But unlike my mother, Ann wasn't afraid. She had chosen to be there, delivering a snapshot of her brain to the hypothalamus of the universe.

Ann says she had prepared a mental itinerary of ideas and figures from world history that she intended to immortalize in the memory of the cosmos. A philosophical and historical report to occupy the hour of recording, something that would make her neurons constellate with noble content and leave a record of thought about humanity. Nevertheless, when she lay down and the examination began, her mind was seized by a very recent memory that took her on a detour from the script. In an act of utter rebellion, her neurons constellated with a will of their own, bypassing those big, important historical ideas and figures. A phone conversation she'd had a few days before crept into her mind. A conversation—most definitely unscientific—with a scientist colleague, a dear friend. A kind of declaration or romantic pact that had shaken her to her foundations and plunged her into a state of profound happiness. Nothing was more powerful than that feeling just then, and so it ended up recorded on the EEG and on the *Voyagers'* golden record, with a sound akin to that of an explosion.

I imagine a group of neurons lighting up in Ann's brain at the memory of that phone conversation. A fluorescent mesh, I guess, strung with the sound of every word, every tone, every breath, every silence. A dizzying sensory array that must have gone off in her hypothalamus like a New Year's Eve fireworks display.

Ann's memory, like the memory of all human beings who ever set foot on this planet, is ungovernable. We try to channel it, marshaling it in every corner of our hypothalamus. If that doesn't work, we resort to other platforms, creating photo albums, recordings, diaries, and every kind of file storable on a hard drive. We need it to be organized, logical, arrayed in story form, unambiguous, and unquestionable so we can find our way, figure ourselves out, understand what we are. But memory is tricky and doesn't follow a script. It was designed that way in our DNA, a hangover from the great explosion at the origin of everything. And this unbridled force makes it move in random ways, without rhyme or reason, in a chaotic, dangerous game. It rears up, raves, curses, demands, shrinks. It attacks us, falling upon us and disarming us unexpectedly. A phone conversation from a few days ago, an enigmatic dream written down in a notebook, a gray hair left on an old cushion, the soft snore of a child, the engine of a helicopter flying over the roof of the house.

Ann's memory, like the memory of all human beings who ever set foot on this planet, follows no rules, has a life of its own, and the record of that rebellion is wandering through space, part of the message sent to the universe.

▶

Beyond photograph and memory, where will we end up? What skill will be needed to pierce the layers of time burying this instant? Where will the laughter of these women go? The smoky smell of blown-out candles, the chocolate crumbs on the white tablecloth? Will they be recycled somehow? Will they turn into dreams? Will they drift lightly like a kite, falling when we least expect it?

▶

I've come to the desert to finish this book. I want to do it by creating a golden record of my own, a personal memory capsule. My first idea was to put a series of things in an envelope and send them up to the stars in a hot-air balloon. Absurd and childish from the start, that project was clearly doomed to fail, so I decided to change tactics and just sit and look up at the night sky, mentally crafting a letter or a thought. Or rather, imagining a space-time capsule to be launched from my brain, traveling as far as it could go.

It's cold. I can feel the icy wind creeping up the sleeves of my jacket. Drowsiness, pent-up fatigue. Soreness in my neck from long minutes of gazing skyward. There is no moon. A vast braid of stars covers the sky tonight. The bright swath of a shining white belt, like the trunk of a great vine sprouting stellar branches and flowers. According to the ancient Greeks, it is milk spilled from the breasts of jealous Hera. In tribute, our galaxy was baptized the Milky Way. In a chapter of the book version of *Cosmos*, Sagan describes how a tribe from Botswana's Kalahari Desert believed the Milky Way was the spine of the body inhabited by all women and men. We were safe in this body because the Milky Way took it upon itself to hold up the weight of the night. If it weren't for those vast, starry branches, pieces of the dark would fall down and break upon us. The Great Spine of Night, they called it.

So many stars are twinkling above me that I can hear them. I listen to their hum, countless voices whispering in my ear. I can imagine what they're saying to me. I'm carrying the whole universe on my shoulders. We all do, all of us. I realized this at my mother's neurological exam. Seeing hundreds of stars sparkling in her brain. The great spine of her own night. Little lamps protecting her from the fearsome darkness.

This will be the first snapshot I put in my capsule. My mother's neurological exam with its record of my beginning. The constellation named for the day I was born. Next I'll add a picture. The one of my mother as a little girl, smiling into the camera, the first photographic record of her. And after that the photograph of Mario that I saw at Violeta's house. And one of Violeta when she was young, maybe twenty or so, standing next to President Salvador Allende on one of his campaigns. Violeta showed it to me proudly, so I'll send it into space to share the sentiment. My son's speech will go too. Mario will enjoy it. So will my grandmother. And I'll include the picture of my grandmother holding my mother in her arms on her birthday in the old house in the port city of San Antonio.

My grandmother was a secretary to the Ministry of Labor. On her old Remington, which now lives on my desk, she recorded every meeting, every speech, every office event. That was her job. Taking notes and keeping records. I saw some of those records typed by her crooked fingers on papers filed in thick ministry folders. Meeting minutes, lists of names, lists of materials, countless letters, resignation papers, welcome speeches, birthday addresses, a true memory capsule of the time she worked there.

I suppose it's from her that I inherited this call to be a space probe, a nosy drone that keeps watch and takes notes. I'm a kind of *Voyager*. Packing less technology, stripped of instruments, cameras, and sensors, functioning with human-scale intelligence that is by now somewhat worse for wear, and equipped with nothing but the few tools of my trade: my worn hypothalamus and my own crooked fingers on the computer keyboard.

What will become of my mother's tired green eyes? What will become of the fine gray hairs she leaves on the armchair cushions? Where will it end up, the laughter of those women, the smoky smell of blown-out candles, the chocolate crumbs on the white tablecloth? Will memory be able to recover it all? Will an exact copy be stored, ready for when we need it? A clear script to keep us from forgetting the voices, the hairstyles, the smells of each body, the lulls in conversation? Will this moment be replayable, at least one more time, in someone else's brain?

The last thing I'll add to my capsule will be this photograph, taken at my mother's eightieth birthday party. This image that is us in the eye of my son's father's camera will be hidden among many others. Maybe we'll end up in a Calpany shoebox or on some computer hard drive. In it we are happy, posing next to a cake, celebrating a woman who was once a little girl and had her picture taken in her mother's arms with other dear faces that no longer exist, in another place long gone, another celebration long past. And in my role as space probe, as recording machine, today I rescue this photograph and I launch it into the stars to drift through time and space, and maybe someday, in some other life, in a future I will never know, someone will find it and choose to pick up the baton of memory.

And whenever someone sees this picture, sparks will be set off in their own brain. A fluorescent mesh strung with familiar, comforting sensory images in which we are the protagonists. We'll be brought back to life in those colors, textures, temperatures, emotions. In the dark room of that head we will live again, twinkling like the lights on a Christmas tree.

Then we'll be a constellation of the future. Broken mirrors of tomorrow's hypothalamus. Tiny people signaling in Morse code, trying to say: *Hello, here we are. Don't forget us.*

Santiago de Chile, May 2019

Nona Fernández was born in Santiago, Chile, in 1971. She is an actress and writer, and has published two plays, a collection of short stories, a work of nonfiction, and six novels, including *The Twilight Zone*, which was a finalist for the National Book Award for Translated Literature, and *Space Invaders*. In 2016 she was awarded the Premio de Literatura Sor Juana Inés de la Cruz. Her books have been translated into French, Italian, German, Greek, Portuguese, Turkish, and English.

Natasha Wimmer is the translator of nine books by Roberto Bolaño, including *The Savage Detectives* and *2666*. Her recent translations include Nona Fernández's *The Twilight Zone* and *Space Invaders*. She lives in Brooklyn with her husband and two children.

The text of *Voyager* is set in Arno Pro.
Book design by Rachel Holscher.
Composition by Bookmobile Design and Digital
Publisher Services, Minneapolis, Minnesota.
Manufactured by Versa Press on acid-free,
30 percent postconsumer wastepaper.